Engineering Mathematics-IV

Engineering Mathematics-IV

Chidanand S. Mujawar
Professor of Mathematics
S.G. Balekundari Institute of Technology, Belgaum
KLE's College of Engineering and Technology, Belgaum
Former Head Dept. of Mathematics
&
Visiting Professor
P.G. Department V.T.U. Belgaum

I.K. International Publishing House Pvt. Ltd.

NEW DELHI • BANGALORE

Published by
I.K. International Publishing House Pvt. Ltd.
S-25, Green Park Extension
Uphaar Cinema Market
New Delhi–110 016 (India)
E-mail : info@ikinternational.com
Website: www.ikbooks.com

ISBN 978-93-81141-68-7

10 9 8 7 6 5 4 3 2 1

Published by Krishan Makhijani for I.K. International Publishing House Pvt. Ltd., S-25, Green Park Extension, Uphaar Cinema Market, New Delhi–110 016 and Printed by Rekha Printers Pvt. Ltd., Okhla Industrial Area, Phase II, New Delhi–110 020.

Dedicated

to

His Holiness

Shri Siddharam Swamiji

Rudrakshimath, Nagnur, Belgaum

"Learning leads to Creativity
Creativity leads to thinking
Thinking provides Knowledge
Knowledge makes you great."

— Dr. A. P. J Abdul Kalam

FOREWORD

Engineering Mathematics-IV authored by Prof. Chidanand S. Mujawar is primarily meant for undergraduate students of Engineering. The presentation in the book is simple, rigorous and compatible with the curriculum of Engineering Mathematics.

This book is self-contained, presentation is detailed with examples and notations used are standard. The chapters are largely independent covering numerical methods, complex variables, special functions, probability theory and sampling theory.

My association with Prof. Mujawar has been a long and a memorable one. He has a tryst with writing mathematics books. His books manifest his deep knowledge of the subject, and because of which, he can explain the complex concepts in a very lucid manner.

The author has been an outstanding teacher over 38 years. He has taught for 34 years Engineering students in KLE Society's College of Engineering and Technology, affiliated to Visvesvaraya Technological University, Belgaum. He deserves our praise and encouragement for accomplishing this task. I wish the readers a good luck and happy reading.

Dr. D.H. Rao
M.E., M.S., Ph.D. (Canada)
MBA, Ph.D. (USA)
Principal & Director
Jain College of Engineering
Belgaum

PREFACE

Engineering Mathematics-IV is designed for the fourth semester of engineering students of Visvesvaraya Technological University. This book is according to the new Syllabus of the University. The book has been divided into eight units, covering the topics Numerical Methods, Complex Variables, Special Functions, Probability Theory and Sampling Theory. Each topic is treated in a systematic and logical manner. An outstanding and distinguishing feature of the book is the large number of typical solved examples followed by well graded question banks for each topic. Many examples and problems have been selected from recent papers of various engineering examinations.
I take this opportunity to thank my colleagues for proper suggestions for the improvement of this book.
I will gratefully accept from the readers any corrections, criticisms and suggestions for the improvement of the book.

Chidanand S. Mujawar

CONTENTS

SYLLABUS

Sub Code	:	10MAT41	IA Marks	:	25
Hrs/Week	:	04	Exam Hours	:	03
Total Hrs.	:	52	Exam Marks	:	100

PART-A

UNIT-1

Numerical Methods – 1

Numerical solution of ordinary differential equations of first order and first degree; Picard's Method Taylor's series method, modified Euler's method, Runge-Kutta method of fourth-order. Milne's and Adams - Bashforth predictor and Corrector Method (No derivations of formulae).

6 Hours

UNIT-2

Numerical Methods – 2

Numerical solution of simultaneous first order ordinary differential equations: Picard's method, Runge-Kutta method of fourth-order.

Numerical solution of second order ordinary differential equations: Picard's method, Runge-Kutta method and Milne's method.

6 Hours

UNIT-3

Complex Variables – 1

Function of a complex variable, Analytic functions-Cauchy-Riemann equations in cartesian and polar forms. Properties of analytic functions.

Application to flow problems complex potential, velocity potential, equipotential lines, stream functions, stream lines.

7 Hours

UNIT-4

Complex Variables – 2

Conformal Transformations: Bilinear Transformations. Discussion of Transformations: $w = z^2$, $w = e^z$, $w = z + (a^2/z)$. Complex line integrals-Cauchy's theorem Cauchy's integral formula.

7 Hours

UNIT-5

Special Functions

Solution of Laplace equation in cylindrical and spherical systems leading Bessel's and Legendre's differential equations, Series solution of Bassel's differential equation leading to Bessel function of first kind. Orthogonal property of Bessel functions. Series solution of Legendre's differential equation leading to Legendre polynomials, Rodrigue's formula.

7 Hours

UNIT-6

Probability Theory – 1

Probability of an event, empherical and axiomatic definition, probability associated with set theory, addition law, conditional probability multiplication law, Baye's theorem.

6 Hours

UNIT-7

Probability Theory – 2

Random variables (discrete and continuous), probability density function, cumulative density function. Probability distributions – Binomial and Poisson distributions; Exponential and normal distributions.

7 Hours

UNIT-8

Sampling Theory

Sampling, Sampling distributions, standard error, test of hypothesis for means, confidence limits for means, student's t-distribution. Chi-Square distribution as a test of goodness of fit.

6 Hours

UNIT 1

Numerical Methods – I

1.1 NUMERICAL SOLUTION OF ORDINARY DIFFERENTIAL EQUATIONS OF THE FIRST ORDER

In this topic, we shall develop certain methods of obtaining the numerical solution to any desired degree of accuracy of first and second order differential equations, consisting of numerical coefficients under given conditions. There are some methods by which numerical solution of any ordinary differential equation having numerical coefficients and the given initial conditions can be found to any desired degree of accuracy. The important methods of solving ordinary differential equations of the first order are as follows:

1. Picard's method of successive approximations.
2. Taylor's series method
3. Modified Euler's method
4. Runge-Kutta method of fourth order
5. Milne's predictor and corrector method
6. Adam's-Bashforth predictor and corrector method.

1.1.1 Picard's Method of Successive Approximations

Let $\frac{dy}{dx} = f(x,y)$ be the initial value problem with the conditions $y = y_0$ at $x = x_0$.

Integrating, this equation between the limits x_0 and x we get

$$\int_{y_0}^{y} dy = \int_{x_0}^{x} f(x,y)\, dx$$

$$\text{or} \quad y - y_0 = \int_{x_0}^{x} f(x,y)dx$$

$$\therefore \quad y = y_0 + \int_{x_0}^{x} f(x,y)dx \tag{1}$$

For first approximation, replace y by y_0 in $f(x,y)$; for second replace it by the first approximation, for a third by the second and so on.

i.e.,

First approximation

$$y^{(1)} = y_0 + \int_{x_0}^{x} f(x,y_0)dx$$

Second approximation

$$y^{(2)} = y_0 + \int_{x_0}^{x} f(x,y^{(1)})dx$$

Third approximation

$$y^{(3)} = y_0 + \int_{x_0}^{x} f(x,y^{(2)})dx$$

....

....

n^{th} approximation

$$y^{(n)} = y_0 + \int_{x_0}^{x} f(x,y^{(n-1)})dx.$$

As a matter of fact the process is stopped when the two values of y, viz., $y^{(n-1)}$ and $y^{(n)}$ are sensibly the same, i.e., they are same to the desired degree of accuracy.

Note: Each approximation gives a better approximation of the required solution than the preceding one.

Example 1: Use Picard's method to approximate y when $x = 0.2$ given that $y = 1$ when $x = 0$ and $\frac{dy}{dx} = x - y$.

Solution: We have

$$f(x,y) = x - y, \quad x_0 = 0, \quad y_0 = 1$$

Now, the first approximation

$$y^{(1)} = y_0 + \int_{x_0}^{x} f(x, y_0)\,dx$$

$$= 1 + \int_{0}^{x} (x-1)\,dx = 1 + \frac{x^2}{2} - x$$

$$= \frac{x^2}{2} - x + 1$$

Second approximation

$$y^{(2)} = y_0 + \int_{x_0}^{x} f(x, y^{(1)})\,dx$$

$$= 1 + \int_{0}^{x} \left(x - \frac{x^2}{2} + x - 1 \right) dx = 1 + \frac{x^2}{2} - \frac{x^3}{6} + \frac{x^2}{2} - x$$

$$= -\frac{x^3}{6} + x^2 - x + 1$$

Third approximation

$$y^{(3)} = y_0 + \int_{x_0}^{x} f(x, y^{(2)})\,dx$$

$$= 1 + \int_{0}^{x} \left(x + \frac{x^3}{6} - x^2 + x - 1 \right) dx$$

$$= \frac{x^4}{24} - \frac{x^3}{3} + x^2 - x + 1$$

Fourth approximation

$$y^{(4)} = y_0 + \int_{x_0}^{x} f(x, y^{(3)})\,dx$$

$$= 1 + \int_0^x \left(x - \frac{x^4}{24} + \frac{x^3}{3} - x^2 + x - 1 \right) dx$$

$$= -\frac{x^5}{120} + \frac{x^4}{12} - \frac{x^3}{3} - x^2 - x + 1$$

When $x = 0.2$

$$y_0 = 1, \quad y^{(1)} = 0.82, \quad y^{(2)} = 0.83862, \quad y^{(3)} = 0.83740, \quad y^{(4)} = 0.83746$$

Thus, when $x = 0.2 \quad y = 0.83746$.

Example 2: Use Picard's method to approximate the value of y when $x = 0.1$ given that $y = 1$ when $x = 0$ and $\frac{dy}{dx} = 3x + y^2$

Solution: We have $f(x,y) = 3x + y^2$, $x_0 = 0$ and $y_0 = 1$. We have to find $y(0.1)$.

First approximation

$$y^{(1)} = y_0 + \int_{x_0}^x f(x, y_0) dx = 1 + \int_{x_0}^x \left(3x + y_0^2\right) dx$$

$$= 1 + \int_{x_0}^x (3x + 1) dx = \frac{3x^2}{2} + x + 1$$

Second approximation

$$y^{(2)} = y_0 + \int_{x_0}^x f(x, y^{(1)}) dx$$

$$= 1 + \int_0^x \left[\frac{9x^4}{4} + 3x^3 + 4x^2 + 5x + 1 \right] dx$$

$$= \frac{9}{20}x^5 + \frac{3}{4}x^4 + \frac{4}{3}x^3 + \frac{5}{2}x^2 + x + 1$$

Third approximation

$$y^{(3)} = y_0 + \int_{x_0}^x f(x, y^{(2)}) dx$$

$$= 1 + \int_0^x \left(\frac{81}{400}x^{10} + \frac{27}{40}x^9 + \frac{141}{80}x^8 + \frac{17}{8}x^7 + \frac{1157}{180}x^6 + \frac{136}{56}x^5 \right.$$

$$\left. + \frac{125}{12}x^4 + \frac{23}{3}x^3 + 6x^2 + 5x + 1 \right] dx$$

$$= \frac{81}{4400}x^{11} + \frac{27}{400}x^{10} + \frac{27}{249}x^9 + \frac{13}{82}x^8 + \frac{1157}{1260}x^7 + \frac{68}{45}x^6 + \frac{25}{11}x^5$$

$$+ \frac{33}{12}x^4 + 2x^3 + \frac{5}{2}x^2 + x + 1.$$

When $x = 0.1$

$$y_0 = 1, \quad y^{(1)} = 1.115, \quad y^{(2)} = 1.1264, \quad y^{(3)} = 1.12721$$

Thus, $y(0.1) = 1.1272$ upto 4 decimal places.

Example 3: Use Picard's method to approximate y when $x = 0.1$ given that $y = 1$ when $x = 0$ and $\frac{dy}{dx} = \frac{y-x}{y+x}$.

Solution: Given $f(x,y) = \frac{y-x}{y+x}$ with $x_0 = 0$, $y_0 = 1$

First approximation

$$y^{(1)} = y_0 + \int_{x_0}^x f(x,y_0)dx = 1 + \int_0^x \frac{1-x}{1+x}dx$$

$$= 1 + \{2\log(1+x) - x\}_0^x = 1 - x + 2\log(1+x)$$

Second approximation

$$y^{(2)} = y_0 + \int_{x_0}^x f(x,y^{(1)})dx$$

$$= 1 + \int_0^x \frac{1 - 2x + 2\log(1+x)}{1 + 2\log(1+x)}dx$$

$$= 1 + x - 2\int_0^x \frac{x}{1 + 2\log(1+x)}dx$$

$$= 1 + x - 2\int_0^t \frac{e^{2t}}{1+2t}dt + 2\int_0^t \frac{e^t}{1+2t}dt \qquad \text{where } t = 1 + 2\log(1+x)$$

Which is difficult to integrate.

$\therefore$ when $x = 0, y = 1$

When $x = 0.1$

$$y^{(1)} = 1 - 0.1 + 2\log(.1)$$

$$= 0.9828$$

$$y^{(2)} = \text{cannot be obtained}$$

Hence, we take the value of y at $x = 0.1$ as $y(0.1) = 0.9828$.

Example 4: Use Picard's method to approximate y when $x = 0.1$, $x = 0.2$, given that $\frac{dy}{dx} = x + y$ with $y = 1$ when $x = 0$. Check the result with exact value.

Solution: Given $f(x,y) = x + y, \quad x_0 = 0, \quad y_0 = 1$

First approximation

$$y^{(1)} = y_0 + \int_{x_0}^x f(x,y_0)dx = 1 + \int_0^x (1+x)dx$$

$$= 1 + x + \frac{x^2}{2}$$

Second approximation

$$y^{(2)} = y_0 + \int_{x_0}^x f(x,y^{(1)})dx$$

$$= 1 + \int_0^x \left[x + \left(\frac{x^2}{2} + x + 1\right)\right] dx$$

$$= \frac{x^3}{6} + x^2 + x + 1$$

Third approximation

$$y^{(3)} = y_0 + \int_0^x \left[x + \left(\frac{x^3}{6} + x^2 + x + 1 \right) \right] dx$$

$$= \frac{x^4}{24} + \frac{x^3}{3} + x^2 + x + 1$$

When $x = 0.1$, we have

$y^{(1)} = 1.105$, $y^{(2)} = 1.11016$, $y^{(3)} = 1.1103$, closer approximation when $x = 0.2$, we have $y^{(3)} = 1.2427$.

Thus, $y(0.1) = 1.1103$, and $y(0.2) = 1.2427$.

Now, we shall obtain the exact value

$$\frac{dy}{dx} = x + y \quad \therefore \quad \frac{dy}{dx} - y = x \quad \text{a linear differential equation.}$$

$$\therefore \quad \text{If} = e^{-\int 1 dx} = e^{-x}$$

$$\therefore \quad ye^{-x} = \int xe^{-x} dx + c$$

$$= x\frac{e^{-x}}{-1} - 1\frac{e^{-x}}{1} + c$$

$$\therefore \quad y = -x - 1 + ce^x$$

Putting $y = 1$, $x = 0$ we get $1 = -1 + c \quad \therefore \quad c = 2$

$\therefore \quad y = -x - 1 + 2e^x$ is the solution.

when $x = 0.1, \quad y = 1.11034$

$x = 0.2, \quad y = 1.2428$

Thus, the results obtained by Picard's method and exact values are upto three decimal places equal.

Example 5: Find the solution of $\frac{dy}{dx} = 1 + xy$ which passes through $(0, 1)$ in the internal $(0, 0.5)$ such that the value of y is correct to 3 decimal places taking $h = 0.1$ by Picard's method.

Solution: Given $f(x,y) = 1 + xy$ with $x_0 = 0, y_0 = 1$.

Now, first approximation

$$y^{(1)} = y_0 + \int_{x_0}^{x} f(x,y_0)dx = 1 + \int_{0}^{x} \{1 + (xy_0)\}\, dx$$

$$= 1 + \int_{0}^{x} (1 + x)dx = 1 + x + \frac{x^2}{2}.$$

Second approximation

$$y^{(2)} = y_0 + \int_{x_0}^{x} f(x,y^{(1)})dx = 1 + \int_{0}^{x} \left\{1 + \left(xy^{(1)}\right)\right\} dx$$

$$= 1 + \int_{0}^{x} \left(1 + x + x^2 + \frac{x^3}{3}\right) dx$$

$$= 1 + x + \frac{x^2}{2} + \frac{x^3}{3} + \frac{x^4}{8}.$$

Third approximation

$$y^{(3)} = 1 + \int_{0}^{x} \left\{1 + \left(xy^{(2)}\right)\right\} dx$$

$$= 1 + \int_{0}^{x} \left\{1 + x\left(1 + x + \frac{x^2}{2} + \frac{x^3}{3} + \frac{x^4}{8}\right)\right\} dx$$

$$= 1 + x + \frac{x^2}{2} + \frac{x^3}{3} + \frac{x^4}{8} + \frac{x^5}{15} + \frac{x^6}{48}.$$

(i) When $x = 0,\quad y = 1$

(ii) When $x = 0.1,\quad y^{(1)} = 1.105,\quad y^{(2)} = 1.10533$

$$\therefore\quad x = 0.1,\quad y = 1.105.$$

(iii) When $x = 0.2,\quad y^{(1)} = 1.202,\quad y^{(2)} = 1.2027$

$$\therefore\quad x = 0.2,\quad y = 1.203.$$

(iv) When $x = 0.3$, $y^{(1)} = 1.345$, $y^{(2)} = 1.35501$, $y^{(3)} = 1.3551$

$$\therefore \quad x = 0.3, \quad y = 1.355.$$

(v) When $x = 0.4$, $y^{(1)} = 1.48$, $y^{(2)} = 1.50453$, $y^{(3)} = 1.5052$

$$\therefore \quad x = 0.4, \quad y = 1.505.$$

(vi) When $x = 0.5$, $y^{(1)} = 1.625$, $y^{(2)} = 1.67447$, $y^{(3)} = 1.6765$

$$\therefore \quad x = 0.5, \quad y = 1.6765.$$

Thus, we can form a table as follows.

x:	0	0.1	0.2	0.3	0.4	0.5
y:	1.000	1.105	1.203	1.355	1.505	1.677

1.1.2 Taylor's Series Method

Let the first order differential equation be $\frac{dy}{dx} = f(x,y)$ with the initial value $y(x_0) = y_0$. Let the solution of this equation be $y(x)$ such that $y(x_0) \neq 0$, then the expansion of $y(x)$, i.e., y in the neighbourhood of x_0 by Taylor's expansion method we obtain

$$y = y_0 + (x - x_0)y_0' + \frac{(x - x_0)^2}{2!} y_0'' + \cdots$$

$$= y_0 + hy_0' + \frac{h^2}{2!} y_0'' + \frac{h^3}{3!} y_0''' + \cdots$$

Which is convergent series in x in the range $x_0 \leq x < x_n$. The coefficients $y_0', y_0'', y_0''', \ldots$ can be obtained by differentiating $y' = f(x,y)$ successively.

Example 1: Using Taylor's series method find y at $x = 0.1, 0.2, 0.3$ (upto 3rd derivative terms) given that $\frac{dy}{dx} = x^2 + y^2$ with $y(0) = 1$

Solution: Given $x_0 = 0$, $y_0 = 1$ and $y' = x^2 + y^2$.

Taylor's series expansion of $y(x)$ is

$$y(x) = y(x_0) + (x - x_0)y'(x_0) + \frac{(x - x_0)^2}{2!} y''(x_0) + \cdots$$

$$\therefore \quad y(x) = y(0) + xy'(0) + \frac{x^2}{2!} y''(0) + \frac{x^3}{3!} y'''(0) + \cdots \qquad (1)$$

Consider

$$y' = x^2 + y^2 \qquad \therefore \quad y'(0) = 1$$

$$y'' = 2x + 2y\,y' \qquad y''(0) = 2 + 1 = 2$$

$$y''' = 2 + 2y'^2 + 2y\,y'' \qquad y'''(0) = 2 + 2 + 2 + 2 = 8 \quad \text{etc.}$$

Substituting in (1)

$$y(x) = 1 + x(1) + \frac{x^2}{2!}(2) + \frac{x^3}{3!}(8) + \cdots$$

$$= 1 + x + x^2 + \frac{4}{3}x^3 + \cdots$$

Now,

$$x = 0.1, \quad y(0.1) = 1 + (.1) + (.1)^2 + \frac{4}{3}(.1)^3 = 1.1113$$

$$x = 0.2, \quad y(0.2) = 1 + (.2) + (.2)^2 + \frac{4}{3}(.2)^3 = 1.2507$$

$$x = 0.3, \quad y(0.3) = 1 + (.3) + (.3)^2 + \frac{4}{3}(.3)^3 = 1.426$$

Example 2: Find y at $x = 1.02$ correct to five decimal places given that $\frac{dy}{dx} = xy - 1$ and $y = 2$ at $x = 1$ applying Taylor's series method.

Solution: Taylor's series expansion is given by

$$y(x) = y(x_0) + (x - x_0)y'(x_0) + \frac{(x - x_0)^2}{2!}y''(x_0) + \cdots$$

Given data is $x_0 = 1$, $y_0 = 2$, and $y' = \frac{dy}{yx} = xy - 1$

$$\therefore \quad y(x) = y(1) + (x - 1)y'(1) + \frac{(x-1)^2}{2!}y''(1) + \frac{(x-1)^3}{3!}y'''(1) + \cdots \qquad (1)$$

Consider

$$y' = xy + 1 \qquad \therefore \quad y'(1) = (1)(2) - 1 = 1$$

$$y'' = xy' + y \qquad y''(1) = (1)(1) + 2 = 3$$

$$y''' = xy'' + y' + y' \qquad y'''(1) = (1)(3) + 1 + 1 = 5 \quad \text{etc.}$$

At $x = 1.02$ and substituting $y'(1), y''(1), y'''(1)\ldots$ in (1)

$$y(1.02) = 2 + (1.02 - 1)(1) + \frac{(1.02-1)^2}{2!}(3) + \frac{(1.02-1)^3}{3!}(5) + \ldots$$

$\therefore \quad y(1.02) = 2.02061.$

Example 3: Using Taylor's series method find $y(0.1)$ considering upto 4th degree term, if $y(x)$ satisfies the equation $\frac{dy}{dx} = x - y^2$ and $y(0) = 1$.

Solution: Taylor's series expansion is given by

$$y(x) = y(x_0) + (x - x_0)y'(x_0) + \frac{(x-x_0)^2}{2!}y''(x_0) + \ldots$$

By data, $x_0 = 0,\ y_0 = 1$ and $y' = x - y^2$

Now,

$$y' = x - y^2 \qquad \therefore \quad y'(0) = 0 - 1^2 = -1$$

$$y'' = 1 - 2yy' \qquad y''(0) = 1 - 2(1)(-1) = 3$$

$$y''' = -2y'^2 - 2yy'' \qquad y'''(0) = -2(-1) - 2(1)(3) = -8$$

$$y^{iv} = -4y'y'' - 2y'y'' - 2yy''' \qquad y^{iv}(0) = -2(-1)(-8) - 2(-1)(3) + -2(1)(-8)$$

$$= 34 \quad \text{etc.}$$

Substituting in (1) $x = 0.1$, and $y'(0), y''(0), y'''(0)$ etc.

we get

$$y(0.1) = 1 + (.1)(-1) + \frac{(0.1)^2}{2!}(3) + \frac{(.1)^3}{3!}(-8) + \frac{(.1)^4}{4!}(34) + \ldots$$

$$= 0.9138$$

Thus, $y(0.1) = 0.9138$.

Example 4: Use Taylor's series method to solve $\frac{dy}{dx} = x^2 + y$ in the range $0 \le x \le 0.2$ by taking step size $h = 0.1$ given that $y = 10$ at $x = 0$ initially considering terms upto the fourth order.

Solution: In this problem, since the step size is 0.1, the problem has to be done in two stages.

Stage I: Given data is $y' = x^2 + y$, $x_0 = 0$ and $y_0 = 10$

Now,

$$\begin{aligned} y' &= x^2 + y \quad \therefore \quad & y'(0) &= 0 + 10 = 10 \\ y'' &= 2x + y' & y''(0) &= 2(0) + 10 = 10 \\ y''' &= 2 + y'' & y'''(0) &= 2 + 10 = 12 \\ y^{iv} &= y''' & y^{iv}(0) &= 12 \quad \text{etc.} \end{aligned}$$

Taylor's series expansion is

$$y(x) = y(x_0) + xy'(x_0) + \frac{x^2}{2!}y''(x_0) + \frac{x^3}{3!}y'''(x_0) + \ldots \qquad (1)$$

$$\therefore \quad y(0.1) = 10 + (.1)(10) + \frac{(.1)^2}{2!}(10) + \frac{(.1)^3}{3!}(12) + \frac{(.1)^4}{4!}(12) + \ldots$$

$$= 11.05205.$$

Stage II: Now taking $x_0 = 0.1$, $y_0 = 11.052$.

$$\begin{aligned} y' &= x^2 + y \quad \therefore \quad & y'(0.1) &= (.1)^2 + 11.052 = 11.062 \\ y'' &= 2x + y' & y''(0.1) &= 2(.1) + (11.062) = 11.262 \\ y''' &= 2x + y'' & y'''(0.1) &= 2 + 11.262 = 13.262 \\ y^{iv} &= y''' & y^{iv}(0.1) &= 13.262 \quad \text{etc.} \end{aligned}$$

$\therefore$ Substituting in Taylor's expansion (1) $x = 0.1$ and $y'(0.1)$, $y''(0.1)$, $y'''(0.1)$ etc., we get

$$y(0.2) = 11.052 + (.1)(11.062) + \frac{(.1)^2}{2!}(11.262) + \frac{(.1)^3}{3!}(13.262) + \ldots$$

$$= 12.216776$$

$$\therefore \quad y(0.2) = 12.2168.$$

Example 5: Use Taylor's series method to obtain approximate value of y at $x = 0.2$ for the differential equation $\frac{dy}{dx} = 2y + 3e^x$, $y(0) = 0$. Compute the exact solution and compare with it.

Solution: Given data is $y' = \frac{dy}{dx} = 2y + 3e^x$; $x_0 = 0$, $y_0 = 0$

Taylor's series expansion about $x_0 = 0$ is

$$y(x) = y(0) + xy'(0) + \frac{x^2}{2!}y''(0) + \frac{x^3}{3!}y'''(0) + \ldots \quad (1)$$

Consider

$$y' = 2y + 3e^x \qquad \therefore \quad y'(0) = 2(0) + 3e^0 = 3$$

$$y'' = 2y' + 3e^x \qquad y''(0) = 2(3) + 3 = 9$$

$$y''' = 2y'' + 3e^x \qquad y'''(0) = 2(9) + 3 = 21$$

$$y^{iv} = 2y''' + 3e^x \qquad y^{iv}(0) = 2(21) + 3 = 45 \quad \text{etc.}$$

Substituting in (1)

$$y(0.2) = 0 + (.2)(3) + \frac{(.2)^2}{2!}(9) + \frac{(.2)^3}{3!}(21) + \frac{(.2)^4}{4!}(45) + \ldots$$

$$= 0.8110$$

$$\therefore \quad y(0.2) = 0.8110 \quad (1)$$

Now,

$$\frac{dy}{dx} = 2y = 3e^x$$

$$\therefore \quad \text{If} = e^{-2\int dx} = e^{-2x}$$

$\therefore$ the solution is

$$ye^{-2x} = \int 3e^x e^{-2x} dx + c$$

$$= -3e^{-x} + c$$

$$\therefore \quad y = -3e^x + ce^{2x}$$

when $x = 0$, $y = 0$ (initial values)

$$\therefore \quad 0 = -3 + c \qquad \therefore \quad c = 3.$$

$$\therefore \quad y = -3e^x + 3e^{2x}.$$

when $x = 0.2$,

$$y = -3e^{.2} + 3e^{2\times.2} = 0.8112 \qquad (2)$$

Comparing (1) and (2) it is clear that the result approximates to the exact value upto 3 decimal places.

Example 6: Use Taylor's Series Method to find $y(4.1)$ given that $\dfrac{dy}{dx} = \dfrac{1}{x^2 + y}$, $y(4) = 4$.

Solution: Given data is $y' = \dfrac{dy}{dx} = \dfrac{1}{x^2 + y}$ with $x_0 = 4$, $y_0 = 4$.

Taylor's series expansion is

$$y(x) = y(x) + (x-4)y'(4) + \frac{(x-4)^2}{2!}y''(4) + \ldots \qquad \text{(i)}$$

Now, $y' = \dfrac{1}{x^2 + y}$

$$\therefore \quad y'(x^2 + y) = 1$$

$$\therefore \quad y'(4)(4^2 + 4) = 1 \qquad \therefore \quad y'(4) = \frac{1}{20}$$

Next

$$y''(x^2 + 4) + y'(2x + 4) = 0$$

$$y'' = \frac{-y'(2x+4)}{(x^2+4)}$$

$$\therefore \quad y''(4) = \frac{-0.4025}{20}$$

$$= -0.02012$$

Substituting in (1)

$$y(4.1) = 4 + (4.1 - 4)\left(\frac{1}{20}\right) + \frac{(4.1-4)^2}{2!}(-0.02012) + \ldots$$

$$= 4.0049$$

$$\therefore \quad y(4.1) = 4.004911.$$

1.1.3 Euler's Method

Consider the initial value problem $\frac{dy}{dx} = f(x,y)$ with $y(x_0) = y_0$.

Denote $x_1 = x_0 + h$, so that Euler's formula becomes

$$y(x_1) = y_1 = y(x_0) + hf\,(x_0, y_0)$$

$$y(x_2) = y_2 = y(x_1) + hf\,(x_1, y_1)$$

$$y(x_3) = y_3 = y(x_2) + hf\,(x_2, y_2)$$

and in general if $x_{n+1} = x_{n+h}$ then we have

$$y(x_{n+1}) = y_{n+1} = y(x_n) + hf\,(x_n, y_n).$$

This is known as Euler's formula.

Modified Euler's Method

Consider the initial value problem $\frac{dy}{dx} = f(x,y)$ with $y(x_0) = y_0$.

We first obtain $y(x_1) = y_1$ by applying Euler's formula

$$y(x_1) = y_1 = y(x_0) + hf(x_0, y_0)$$

This value is regarded as the initial approximation for y_1 usually denoted by $y_1^{(0)}$. It is also called the predicated value of y_1.

$$\therefore \quad y_1^{(0)} = y_0 + hf(x_0, y_0). \quad \text{(Euler's formula)}$$

This value is successively improved to desired degree of accuracy by the modified Euler's formula where the successive approximations are denoted by $y_1^{(1)}$, $y_1^{(2)}$, $y_1^{(3)}$, etc.

$$y_1^{(1)} = y_0 + \frac{h}{2}\left[f(x_0, y_0) + f(x_1, y_1^{(0)})\right]$$

$$y_1^{(2)} = y_0 + \frac{h}{2}\left[f(x_0, y_0) + f(x_1, y_1^{(1)})\right]$$

$$y_1^{(3)} = y_0 + \frac{h}{2}\left[f(x_0, y_0) + f(x_1, y_1^{(2)})\right]$$

$$y_1^{(4)} = y_0 + \frac{h}{2}\left[f(x_0, y_0) + f(x_1, y_1^{(3)})\right].$$

and so on.

Each of the succeeding approximation is better than the preceding ones and are also called the corrected value.

Next,

$$y_2^{(0)} = y_1 + hf(x_1, y_1) \quad \text{(Euler's formula)}$$

$$y_2^{(1)} = y_1 + h\left[f(x_1, y_1) + f(x_2, y_2^{(0)})\right]$$

$$y_2^{(2)} = y_1 + h\left[f(x_1, y_1) + f(x_2, y_2^{(1)})\right]$$

$$y_2^{(3)} = y_1 + h\left[f(x_1, y_1) + f(x_2, y_2^{(2)})\right]$$

and so on.

Euler's formula and modified Euler's formula jointly called Euler's predictor and corrector formulae.

Example 1: Using Modified Euler's Method find $y(0.2)$ correct to four decimal places solving the differential equation $\frac{dy}{dx} = x - y^2$, $y(0) = 1$. Taking $h = 0.1$.

Solution:

First Stage:

By data $f(x, y) = x - y^2$, $x_0 = 0$, $y_0 = 1$ and $h = 0.1$

$$\therefore \quad f(x_0, y_0) = 0 - 1^2 = -1. \quad x_1 = x_0 + h = 0 + 0.1 = 0.1$$

From Euler's formula

$$y_1^{(0)} = y_0 + hf(x_0, y_0)$$

$$= 1 + (.1)[0 - 1^2] = 0.9.$$

By modified Euler's formula

$$y_1^{(1)} = y_0 + \frac{h}{2}\left[f(x_0, y_0) + f(x_1\ y_1^{(0)})\right]$$

$$= 1 + \frac{0.1}{2}\left[(-1) + (.1 - (.9)^2)\right] = 0.9145$$

$$y_1^{(2)} = y_0 + \frac{h}{2}\left[f(x_0, y_0) + f(x_1\ y_1^{(1)})\right]$$

$$= 1 + \frac{0.1}{2}\left[(-1) + (.1 - (0.9145)^2)\right] = 0.9132$$

$$y_1^{(3)} = y_0 + \frac{h}{2}\left[f(x_0, y_0) + f(x_1\ y_1^{(2)})\right]$$

$$= 1 + \frac{0.1}{2}\left[(-1) + (.1 - (0.9132)^2)\right] = 0.9133$$

$$\therefore \quad y(0.1) = 0.9133$$

Second Stage:

Now take $x_1 = 0.1\ y_1 = 0.9133$ and $f(x,y) = x - y^2$

$$\therefore \quad f(x_1, y_1) = 0.1 - (0.9133)^2 = -0.7341$$

$$x_2 = x_1 + h = 0.2 \quad y(x_2) = y_2 = ?$$

By Euler's formula

$$y_2^{(0)} = y_1 + hf(x_1\ y_1)$$

$$= 0.9133 + 0.1(-0.7341) = 0.8399$$

By modified Euler's formula

$$y_2^{(1)} = y_1 + \frac{h}{2}\left[f(x_1 y_1) + f(x_2\ y_2^{(0)})\right]$$

$$= 0.9133 + \frac{.1}{2}\left[-0.7341 + (.2 - (0.8399)^2)\right]$$

$$= 0.8513$$

$$y_2^{(2)} = y_1 + \frac{h}{2}\left[f(x_1 y_1) + f(x_2\ y_2^{(1)})\right]$$

$$= 0.9133 + \frac{0.1}{2}\left[-0.7341 + (0.2 + (0.8513)^2)\right]$$

$$= 0.8504$$

$$y_2^{(3)} = y_1 + \frac{h}{2}\left[f(x_1 y_1) + f(x_2\ y_2^{(2)})\right]$$

$$= 0.9133 + \frac{0.1}{2}\left[-0.7341 + (0.2 + (0.8504)^2)\right]$$

$$= 0.8504$$

$$\therefore \quad y(0.2) = 0.8504.$$

Example 2: Given $\frac{dy}{dx} = 1 + \frac{y}{x}$, $y = 2$ at $x = 1$. Find approximate value of y at $x = 1.4$ by taking step size $h = 0.2$ apply modified Euler's Method.

Solution:

First Stage:

Here $x_0 = 1$, $y_0 = 2$, $h = 0.2$ and $f(x,y) = 1 + \frac{y}{x}$.

By Euler's formula

$$y_1^{(0)} = y_0 + hf(x_0\ y_0)$$

$$= 2 + (0.2)\left(1 + \frac{2}{1}\right) = 2.6$$

By modified Euler's formula

$$y_1^{(1)} = y_0 + \frac{h}{2}\left[f(x_0, y_0) + f(x_1\ y_1^{(0)})\right]$$

$$= 2 + \frac{0.2}{2}\left[\left(1 + \frac{2}{1}\right) + \left(1 + \frac{2.6}{1.2}\right)\right]$$

$$= 2.6167$$

$$y_1^{(2)} = y_0 + \frac{h}{2}\left[f(x_0, y_0) + f(x_1\ y_1^{(1)})\right]$$

$$= 2 + \frac{0.2}{2}\left[\left(1 + \frac{2}{1}\right) + \left(1 + \frac{2.6167}{1.2}\right)\right]$$

$$= 2.6181$$

$$y_1^{(3)} = y_0 + \frac{h}{2}\left[f(x_0, y_0) + f(x_1\ y_1^{(2)})\right]$$

$$= 2 + \frac{0.2}{2}\left[\left(1 + \frac{2}{1}\right) + \left(1 + \frac{2.6168}{1.2}\right)\right]$$

$$= 2.6182$$

Thus, $y(1.2) = 2.6182.$

Second Stage:

Take $x_1 = 1.2,\ y_1 = 2.6182$

$\therefore \quad f(x_1, y_1) = 1 + \frac{y_1}{x_1}$

$x_2 = x_1 + h = 1.4$, then $y(x_2) = ?$

By Euler's formula we have

$$y_2^{(0)} = y_1 + hf(x_1\ y_1)$$

$$= 2.6182 + (0.2)\left[1 + \frac{2.6182}{1.2}\right]$$

$$= 3.2546$$

By modified Euler's formula

$$y_2^{(1)} = y_1 + \frac{h}{2}\left[f(x_1, y_1) + f(x_2\ y_2^{(0)})\right]$$

$$= 2.6182 + \frac{0.2}{2}\left[\left(1 + \frac{2.6182}{1.2}\right) + \left(1 + \frac{3.2546}{1.4}\right)\right]$$

$$= 3.2689$$

$$y_2^{(2)} = y_1 + \frac{h}{2}\left[f(x_1, y_1) + f(x_2\ y_2^{(1)})\right]$$

$$= 2.6182 + \frac{0.2}{2}\left[\left(1 + \frac{2.6182}{1.2}\right) + \left(1 + \frac{3.2689}{1.4}\right)\right]$$

$$= 3.2699$$

$$y_2^{(3)} = y_1 + \frac{h}{2}\left[f(x_1, y_1) + f(x_2\ y_2^{(2)})\right]$$

$$= 2.6182 + \frac{0.2}{2}\left[\left(1 + \frac{2.6182}{1.2}\right) + \left(1 + \frac{3.2699}{1.4}\right)\right]$$

$$= 3.2699$$

Thus, $y(1.4) = 3.2699.$

Example 3: Use modified Euler's method to solve $\frac{dy}{dx} = x + |\sqrt{y}|$ in the range $0 \le x \le 0.4$ by taking $h = 0.2$ given that $y = 1$ at $x = 0$ initially.

Solution: **First Stage:**

$x_0 = 0,\ y_0 = 1,\ h = 0.2$ and $f(x,y) = x + \sqrt{y}$

$x_1 = x_0 + h = 0 + 0.2 = 0.2 \quad y(0.2) = ?$

By Euler's method

$$y_1^{(0)} = y_0 + hf(x_0\, y_0)$$
$$= 1 + 0(.2)\left[0 + \sqrt{1}\right] = 1.2$$

By modified Euler's method

$$y_1^{(1)} = y_0 + \frac{h}{2}\left[f(x_0, y_0) + f(x_1\, y_1^{(0)})\right]$$
$$= 1 + \frac{0.2}{2}\left[\left(0 + \sqrt{1}\right) + \left(0.2 + \sqrt{1.2}\right)\right]$$
$$= 1.2295$$
$$y_1^{(2)} = y_0 + \frac{h}{2}\left[f(x_0, y_0) + f(x_1\, y_1^{(1)})\right]$$
$$= 1 + \frac{0.2}{2}\left[\left(0 + \sqrt{1}\right) + \left(0.2 + \sqrt{1.2295}\right)\right]$$
$$= 1.2309$$
$$y_1^{(3)} = y_0 + \frac{h}{2}\left[f(x_0, y_0) + f(x_1\, y_1^{(2)})\right]$$
$$= 1 + \frac{0.2}{2}\left[\left(0 + \sqrt{1}\right) + \left(0.2 + \sqrt{1.2309}\right)\right]$$
$$= 1.2309$$

Thus, $y(0.2) = 1.2309$.

Second Stage:

Here $x_1 = 0.2,\ y_1 = 1.2309,\ h = 0.2$

$x_2 = x_1 + h = 0.2 + 0.2 = 0.4, y(x_2) = y(0.4) = ?$

By Euler's formula

$$y_2^{(0)} = y_1 + hf(x_1\, y_1)$$
$$= 1.2309 + 0.2\left(0.2 + \sqrt{1.2309}\right)$$
$$= 1.4927$$

By modified Euler's method

$$y_2^{(1)} = y_1 + \frac{h}{2}\left[f(x_1, y_1) + f(x_2\, y_2^{(0)})\right]$$
$$= 1.2309 + \frac{0.2}{2}\left[\left(0.2 + \sqrt{1.2309}\right) + \left(0.4 + \sqrt{1.4929}\right)\right]$$
$$= 1.5240$$

$$y_2^{(2)} = y_1 + \frac{h}{2}\left[f(x_1, y_1) + f(x_2\, y_2^{(1)})\right]$$
$$= 1.2309 + \frac{0.2}{2}\left[\left(0.2 + \sqrt{1.2309}\right) + \left(0.4 + \sqrt{1.5240}\right)\right]$$
$$= 1.5253$$

$$y_2^{(3)} = y_1 + \frac{h}{2}\left[f(x_1, y_1) + f(x_2\, y_2^{(2)})\right]$$
$$= 1.2309 + \frac{0.2}{2}\left[\left(0.2 + \sqrt{1.2309}\right) + \left(0.4 + \sqrt{1.5253}\right)\right]$$
$$= 1.5253$$

Thus, $y(0.4) = 1.5253$.

Example 4: Given that $\frac{dy}{dx} = \log(x + y)$ with the initial condition that $y = 1$ when $x = 0$, find y for $x = 0.2$ and $x = 0.5$.

Solution: **First Stage:**

Here $x_0 = 0$, $y_0 = 1$, $h = 0.2$ and $f(x, y) = \log(x + y)$.

$x_1 = x_0 + h = 0 + 0.2 = 0.2$, $\quad y(x_1) = y(0.2) = ?$

By Euler's formula

$$y_1^{(0)} = y_0 + hf(x_0\, y_0)$$
$$= 1 + 0.2\,[\log(0 + 1)] = 1$$

By modified Euler's method

$$y_1^{(1)} = y_0 + \frac{h}{2}\left[f(x_0, y_0) + f(x_1\, y_1^{(0)})\right]$$

$$= 1 + \frac{0.2}{2}[\log(0+1) + \log(0.2+1)] \qquad (\log 1 = 0)$$

$$= 1.0079$$

$$y_1^{(2)} = y_0 + \frac{h}{2}\left[f(x_0, y_0) + f(x_1\, y_1^{(1)})\right]$$

$$= 1 + \frac{0.2}{2}[\log 1 + \log(0.2+1.0079)]$$

$$= 1.0082$$

$$y_1^{(3)} = y_0 + \frac{h}{2}\left[f(x_0, y_0) + f(x_1\, y_1^{(2)})\right]$$

$$= 1 + \frac{0.2}{2}[\log 1 + \log(0.2+1.0082)]$$

$$= 1.0082$$

Thus,

$\therefore \quad y(0.2) = 1.0082.$

Second Stage:

Here $x_1 = 0.2$, $y_1 = 1.0082$, $h = 0.5 - 0.2 = 0.3$

and $f(x,y) = \log(x+y)$

$x_2 = x_1 + h = 0.2 + 0.3 = 0.5, \quad \therefore \quad y(x_2) = y(0.5) = ?$

By Euler's formula

$$y_2^{(0)} = y_1 + hf(x_1\, y_1)$$

$$= 1.0082 + 0.3\,[\log(0.2+1.0082)]$$

$$= 1.0328$$

By modified Euler's formula

$$y_2^{(1)} = y_1 + \frac{h}{2}\left[f(x_1, y_1) + f(x_2\, y_2^{(0)})\right]$$

$$= 1.0082 + \frac{0.3}{2} [\log(0.2 + 1.0082) + \log(0.5 + 1.0328)]$$

$$= 1.0483$$

$$y_2^{(2)} = y_1 + \frac{h}{2}\left[f(x_1, y_1) + f(x_2\, y_2^{(1)})\right]$$

$$= 1.0082 + \frac{0.3}{2} [\log(0.2 + 1.0082) + \log(0.5 + 1.0483)]$$

$$= 1.0490$$

$$y_2^{(3)} = y_1 + \frac{h}{2}\left[f(x_1, y_1) + f(x_2\, y_2^{(2)})\right]$$

$$= 1.0490$$

Thus, $y(0.5) = 1.0490$.

Example 5: Given that $\frac{dy}{dx} = x + y$ with initial condition that $y = 1$ when $x = 0$ for $x = 0.05$ and $x = 0.1$

Solution: **First Stage:**

Here $x_0 = 0$, $y_0 = 1$, $h = 0.05$ and $f(x, y) = x + y$.

$x_1 = x_0 + h = 0 + 0.05 = 0.05$.

By Euler's formula

$$y_1^{(0)} = y_0 + hf(x_0\, y_0)$$

$$= 1 + 0.05\,[0 + 1] = 1.05.$$

By modified Euler's method

$$y_1^{(1)} = y_0 + \frac{h}{2}\left[f(x_0, y_0) + f(x_1\, y_1^{(0)})\right]$$

$$= 1 + \frac{0.05}{2} [(0 + 1) + (0.05 + 1.05)]$$

$$= 1.1025$$

$$y_1^{(2)} = y_0 + \frac{h}{2}\left[f(x_0, y_0) + f(x_1\, y_1^{(1)})\right]$$

$$= 1 + \frac{0.05}{2}[(0+1)+(0.05+1.1025)]$$

$$= 1.05256$$

$$y_1^{(3)} = y_0 + \frac{h}{2}\left[f(x_0, y_0) + f(x_1\, y_1^{(2)})\right]$$

$$= 1 + \frac{0.05}{2}[(0+1)+(0.05+1.05256)]$$

$$= 1.05256$$

Thus, $y(0.05) = 1.05256.$

Second Stage:

Take $x_1 = 0.05,\ y_1 = 1.05256,\ h = 0.05$

$f(x, y) = x + y$

$x_2 = x_1 + h = 0.05 + 0.05 = 0.1 \quad \therefore \quad y(0.1) = ?$

By Euler's formula

$$y_2^{(0)} = y_1 + hf(x_0\, y_0)$$

$$= 1.05256 + 0.05\,[0.05 + 1.05256]$$

$$= 1.1103$$

By modified Euler's method

$$y_2^{(1)} = y_1 + \frac{h}{2}\left[f(x_1, y_1) + f(x_2\, y_2^{(0)})\right]$$

$$= 1.05256 + \frac{0.05}{2}[(0.05 + 1.05256) + (0.1 + 1.1103)]$$

$$= 1.1104$$

$$y_2^{(2)} = y_1 + \frac{h}{2}\left[f(x_1, y_1) + f(x_2\, y_2^{(1)})\right]$$

$$= 1.05256 + \frac{0.05}{2}[(0.05 + 1.05256) + (0.1 + 1.1104)]$$

$$= 1.1104$$

Thus, $y(0.1) = 1.1104.$

1.1.4 Runge-Kutta Method

Consider the initial value problem $\frac{dy}{dx} = f(x,y)$ with $y(x_0) = y_0$. Now, we need to find $y(x_0 + h)$ where h is the step size.

Type I: Second order Runge-Kutta method.

We compute k_1 and k_2 as

$$k_1 = hf(x_0, y_0)$$

$$k_2 = hf\left(x_0 + \frac{h}{2}, y_0 + \frac{k_1}{2}\right)$$

$$\therefore \quad y(x_0 + h) = y_0 + k_2$$

Type II: Fourth order Runge-Kutta method.

We compute k_1, k_2, k_3 and k_4 as

$$k_1 = hf(x_0, y_0)$$

$$k_2 = hf\left(x_0 + \frac{h}{2}, y_0 + \frac{k_1}{2}\right)$$

$$k_3 = hf\left(x_0 + \frac{h}{2}, y_0 + \frac{k_2}{2}\right)$$

$$k_4 = hf(x_0 + h, y_0 + k_3)$$

Then $k = \frac{1}{6}(k_1 + 2k_2 + 2k_3 + k_4)$

$$\therefore \quad y(x_0 + h) = y_0 + k.$$

Example 1: Given $\frac{dy}{dx} = 3x + \frac{y}{2}$, $y(0) = 1$ compute $y(0.2)$ by taking $h = 0.2$, using Runge-Kutta fourth order method.

Solution: By data $f(x,y) = 3x + \frac{y}{2}$, $x_0 = 0$, $y_0 = 1$ and $h = 0.2$

We compute k_1, k_2, k_3 and k_4 as follows

$$k_1 = hf(x_0, y_0) = 0.2\left[3(0) + \frac{1}{2}\right] = 0.1$$

$$k_2 = hf\left(x_0 + \frac{h}{2}, y_0 + \frac{k_1}{2}\right)$$

$$= 0.2\left[3\left(0 + \frac{.2}{2}\right) + \frac{\left(1 + \frac{.1}{2}\right)}{2}\right]$$

$$= 0.165$$

$$k_3 = hf\left(x_0 + \frac{h}{2}, y_0 + \frac{k_2}{2}\right)$$

$$= 0.2\left[3\left(0 + \frac{.2}{2}\right) + \frac{\left(1 + \frac{0.165}{2}\right)}{2}\right]$$

$$= 0.16825$$

$$k_4 = hf\left(x_0 + h, y_0 + k_3\right)$$

$$= 0.2\left[3(0 + 0.2) + \frac{(1 + 0.16825)}{2}\right]$$

$$= 0.236825$$

Then $k = \frac{1}{6}(k_1 + 2k_2 + 2k_3 + k_4)$

$$= \frac{1}{6}[0.1 + 2(0.165) + 2(0.16825) + 0.236825]$$

$$= 0.1672208$$

$$\therefore \quad y(0.2) = y_0 + k = 1 + 0.1672208$$

$$= 1.1672208.$$

Example 2: Use fourth order Runge-Kutta method to solve $(x + y)\frac{dy}{dx} = 1$, $y(0.4) = 1$ at $x = 0.5$ correct to four decimal places.

Solution: Given $f(x,y) = \frac{1}{x + y}$, $x_0 = 0.4$, $y_0 = 1$ and $h = 0.1$

We compute k_1, k_2, k_3 and k_4 as follows

$$k_1 = hf(x_0, y_0)$$

$$= 0.1\left[\frac{1}{0.4+1}\right] = 0.0714$$

$$k_2 = hf\left(x_0 + \frac{h}{2}, y_0 + \frac{k_1}{2}\right)$$

$$= 0.1\left[\frac{1}{\left(0.4+\frac{.1}{2}\right)+\left(1+\frac{0.0714}{2}\right)}\right] = 0.0673$$

$$k_3 = hf\left(x_0 + \frac{h}{2}, y_0 + \frac{k_2}{2}\right)$$

$$= 0.1\left[\frac{1}{\left(0.4+\frac{0.11}{2}\right)+\left(1+\frac{0.0674}{2}\right)}\right] = 0.0674$$

$$k_4 = hf(x_0 + h, y_0 + k_3)$$

$$= 0.1\left[\frac{1}{(0.4+.1)+(1+0.0638)}\right] = 0.0638$$

$$\therefore \quad y(x_0 + h) = y_0 + \frac{1}{6}(k_1 + 2k_2 + 2k_3 + k_4)$$

$$= 1 + 0.0674$$

$$= 1.0674$$

$$\therefore \quad y(0.5) = 1.0674.$$

Example 3: Use fourth order Runge-Kutta method to find y at $x = 0.1$ given that $\frac{dy}{dx} = 3e^x + 2y$, $y(0) = 0$ and $h = 0.1$.

Solution: By data $f(x,y) = 3e^x + 2y$, $x_0 = 0$, $y_0 = 0$ and $h = 0.1$ we compute k_1, k_2, k_3 and k_4 as follows

$$K_1 = hf(x_0\ y_0)$$

$$= 0.1\left[3e^0 + 2(0)\right] = 0.3$$

$$K_2 = hf(x_0 + \frac{h}{2}, y_0 + \frac{k_1}{2})$$

$$= 0.1\left[3e^{\left(0+\frac{.1}{2}\right)} + \left(0 + \frac{.3}{2}\right)\right] = 0.3454$$

$$K_3 = hf\left(x_0 + \frac{h}{2}, y_0 + \frac{k_2}{2}\right)$$

$$= 0.1\left[3\,e^{\left(0+\frac{.1}{2}\right)} + \left(0 + \frac{0.3454}{2}\right)\right] = 0.3499$$

$$K_4 = hf(x_0 + h, y_0 + h_3)$$

$$= 0.1\left[3e^{(0+.1)} + (0 + 0.3499)\right] = 0.4015$$

$$\therefore \quad y(x_0 + h) = y_0 + \frac{1}{6}(k_1 + 2k_2 + 2k_3 + k_4)$$

$$= 0 + 0.3487$$

$$= 0.3487$$

Thus, $y(0.1) = 0.3487$

Example 4: Using Runge-Kutta method of order four find y for $x = 0.1$ and $x = 0.2$ given that $\frac{dy}{dx} = x + y^2$ with $y(0) = 1$

Solution: By data $f(x,y) = x + y^2$, $x_0 = 0$, $y_0 = 1$ and $h = 0.1$

Step 1: Take $x_0 = 0$, $y_0 = 1$, $h = 0.1$ and $f(xy) = x + y^2$.

we compute,

$$K_1 = hf(x_0, y_0)$$

$$= 0.1\left[0 + (1)^2\right] = 0.1$$

$$K_2 = hf\left(x_0 + \frac{h}{2}, y_0 + \frac{k_1}{2}\right)$$

$$= 0.1\left[\left(0+\frac{.1}{2}\right)+\left(1+\frac{.1}{2}\right)^2\right] = 0.1152$$

$$K_3 = hf\left(x_0+\frac{h}{2}, y_0+\frac{k_2}{2}\right)$$

$$= 0.1\left[\left(0+\frac{.1}{2}\right)+\left(1+\frac{0.1152}{2}\right)^2\right] = 0.1168$$

$$K_4 = hf\left(x_0+h, y_0+k_3\right)$$

$$= 0.1\left[(0+.1)+(1+0.1168)^2\right] = 0.1347$$

$$\therefore \quad y(x_0+h) = y_0+\frac{1}{6}\left(k_1+2k_2+2k_3+k_4\right)$$

$$= 1+0.1165$$

$$\therefore \quad y(0.1) = 1.1165$$

Step 2: Now $x_1 = x_0 + h = 0.1$, $y_1 = 1.1165$, $h = 0.1$ and $f(xy) = x + y^2$

we compute

$$K_1 = hf(x_1\, y_1) = 0.1\left[0.1+(1.1165)^2\right] = 0.1347$$

$$K_2 = hf\left(x_1+\frac{h}{2}, y_1+\frac{k_1}{2}\right)$$

$$= 0.1\left[\left(0.1+\frac{0.1}{2}\right)+\left(1.1165+\frac{0.1347}{2}\right)^2\right]$$

$$= 0.1551$$

$$K_3 = hf\left(x_1+\frac{h}{2}, y_1+\frac{k_2}{2}\right)$$

$$= 0.1\left[\left(0.1+\frac{0.1}{2}\right)+\left(1.1165+\frac{0.1551}{2}\right)^2\right]$$

$$= 0.1576$$

$$K_4 = hf\left(x_1 + h, y_1 + k_3\right)$$
$$= 0.\left[(0.1 + 0.1) + (1.1165 + 0.1576)^2\right]$$
$$= 0.1823$$

$$\therefore \quad y(x_1 + h) = y_1 + \frac{1}{6}\left(k_1 + 2k_2 + 2k_3 + k_4\right)$$
$$\therefore \quad y(.1 + .1) = 1.1165 + 0.1571$$
$$= 1.2736$$
$$\therefore \quad y(0.2) = 1.2736$$

1.1.5 Milne's Predictor and Corrector Formulae

Let $\frac{dy}{dx} = f(x,y)$ be the given differential equation with a set of four pre-determined values of y.

$$y(x_0) = y_0,\ y(x_0 + h) = y_1, \quad y(x_0 + 2h) = y_2 \text{ and } y(x_0 + 3h) = y_3$$

we find the expression for $y(x_0 + 4h) = y_4$.

The predictor and corrector formulae can be written as

$$y_{n+1}^{(P)} = y_{n-3} + \frac{4h}{3}\left(2y'_{n-2} - y'_{n-1} + 2y'_n\right) \tag{1}$$

and

$$y_{n+1}^{(c)} = y_{n-1} + \frac{h}{3}\left(y'_{n-1} + 4y'_n + y'_{n+1}\right) \tag{2}$$

when $n = 3$ we write

$$y_4^{(P)} = y_0 + \frac{4h}{3}(2y'_1 - y'_2 + 2y'_3) \tag{3}$$

$$\text{and } y_4^{(c)} = y_2 + \frac{h}{3}(y'_2 + 4y'_3 + y'_4) \tag{4}$$

(3) is called Milne's predictor formula and (4) is called Milne's corrector formula.

Example 1: Given $\frac{dy}{dx} = x - y^2$ and the data

x:	0.0	0.2	0.4	0.6
y:	0.0	0.02	0.0795	0.1762

Compute y at $x = 0.8$ by Milne's method.

Solution: Given differential equation is

$$\frac{dy}{dx} = x - y^2$$

we shall prepare a table as follows.

x	y	$y^1 = x - y^2$
$x_0 = 0$	$y_0 = 0$	$y_0^1 = 0$
$x_1 = 0.2$	$y_1 = 0.02$	$y_1^1 = .2 - (.02)^2 = 0.1996$
$x_2 = 0.4$	$y_2 = 0.0759$	$y_2^1 = .4 - (.0795)^2 = 0.3937$
$x_3 = 0.6$	$y_3 = 0.1762$	$y_3^1 = 0.6 - (.01762)^2 = 0.5689.$
$x_4 = 0.8$	$y_4 = ?$	

$$y_4^{(p)} = y_0 + \frac{4h}{3}(2y_1^1 - y_2^1 + 2y_3^1)$$

$$= 0 + \frac{4}{3}(0.2)\,[2 \times 0.1996 - 0.3937 + 2 \times 0.5689]$$

$$= 0.3049$$

Now, $y_4^1 = x_4 - y_4^2 = 0.8 - (0.3049)^2 = 0.707$

Next

$$y_4^{(c)} = y_2 + \frac{h}{3}\left[y_2^1 + 4y_3^1 + y_4^1\right]$$

$$= 0.0759 + \frac{0.2}{3}[0.3937 + 4 \times 0.5689 + 0.707]$$

$$= 0.3046$$

Again, $y_4^1 = x_4 - (y_4^{(c)})^2 = 0.8 - (0.3046)^2$

$$= 0.7072$$

$$\therefore \quad y_4^{(c)} = y_2 + \frac{h}{3}\left[y_2^1 + 4y_3^1 + y_4^1\right]$$

$$= 0.0759 + \frac{0.2}{3}[0.3937 + 4 \times 0.5689 + 0.7072]$$

$$= 0.3046$$

Thus, $y_4 = 0.3046$ at $x = 0.8$.

Example 2: Apply Milne's method to compute $y(1.4)$ correct to four decimal places given $\frac{dy}{dx} = x^2 + \frac{y}{2}$ and the data $y(1) = 2$, $y(1.1) = 2.2156$, $y(1.2) = 2.4649$, $y(1.3) = 2.7514$.

Solution: Given the differential equation

$$\frac{dy}{dx} = x^2 + \frac{y}{2}.$$

First we shall prepare the table as follows.

x	y	$y^1 = x^2 + \frac{y}{2}$
$x_0 = 1$	$y_0 = 2$	$y_0^1 = 1^2 + \frac{2}{2} = 2$
$x_1 = 1.1$	$y_1 = 2.2156$	$y_1^1 = (1.1)^2 + \frac{2.2156}{2} = 2.3178$
$x_2 = 1.2$	$y_2 = 2.4649$	$y_2^1 = (1.2)^2 + \frac{2.4649}{2} = 2.67245$
$x_3 = 1.3$	$y_3 = 2.7514$	$y_3^1 = (1.3)^2 + \frac{2.7514}{2} = 3.0657.$
$x_4 = 1.4$	$y_4 = ?$	

we have,

$$y_4^{(P)} = y_0 + \frac{4h}{3}(2y_1^1 - y_2^1 + 2y_3^1)$$

$$= 2 + \frac{4(0.1)}{3}[2 \times (2.3178) - 2.4649 + 2(3.0657)]$$

$$= 3.0793.$$

Now, $y_4^1 = x_4^2 + \frac{y_4}{2} = (1.4)^2 + \frac{(3.0793)}{2} = 3.49965$

$$\therefore\ y_4^{(c)} = y_2 + \frac{h}{3}\left(y_2^1 + 4y_3^1 + y_4^1\right)$$

$$= 2.4649 + \frac{(0.1)}{3}[2.67245 + 4(3.0657) + 3.49965]$$

$$= 3.0794.$$

Again

$$y_4^1 = x_4^2 + \frac{y_4^{(c)}}{2} = (1.4)^2 + \frac{3.0794}{2} = 3.4997$$

$$\therefore \quad y_4^{(c)} = y_2 + \frac{h}{3}\left(y_2^1 + 4y_3^1 + y_4^1\right)$$

$$= 2.4649 + \frac{(0.1)}{3}[2.67245 + 4(3.0657) + 3.4997]$$

$$= 3.0794.$$

$$\therefore \quad y_4^{(c)} = 3.0794$$

Thus, when $x = 1.4,\ y = 3.0794$

Example 3: Use Taylor's Series method to find y at $x = 0.1,\ 0.2,\ 0.3$ (upto third derivatives) given $\frac{dy}{dx} = x^2 + y^2$ with $y(0) = 1$. Apply Milne's predictor-corrector formula to find $y(0.4)$.

Solution: First we use Taylor's series expansion to find $x = 0.1,\ x = 0.2$ and $x = 0.3$. Since $x_0 = 0$, Taylor's expansion is

$$y(x) = y(0) + xy'(0) + \frac{x^2}{2!}y''(0) + \frac{x^3}{3!}y'''(0) + \ldots$$

By data $x_0 = 0,\ y_0 = 1,\ y' = x^2 + y^2$

$$\therefore \qquad\qquad \therefore \quad y'(0) = 0 + 1 = 1$$

$$y'' = 2x + 2y\,y' \qquad y''(0) = 0 + 2 \times 1 \times 1 = 2$$

$$y''' = 2 + 2y^{12} + 2y\,y'' \qquad y'''(0) = 2 + 2 \times 1 + 2 \times 1 \times 2 = 8$$

Substituting in Taylor's expansion

$$y(x) = 1 + x(1) + \frac{x^2}{2!}(2) + \frac{x^3}{3!}(8) + \ldots$$

$\therefore \quad y(0) = 1 \quad y(0.1) = 1.1113, \quad y(0.2) = 1.2507, \quad y(0.3) = 1.426$

We prepare the corresponding table as follows

x	y	$y^1 = x^2 + y^2$
$x_0 = 0$	$y_0 = 1$	$y_0^1 = 1$
$x_1 = 0.1$	$y_1 = 1.1113$	$y_1^1 = 1.245$
$x_2 = 0.2$	$y_2 = 1.2507$	$y_2^1 = 1.6043$
$x_3 = 0.3$	$y_3 = 1.426$	$y_3^1 = 2.1235$
$x_4 = 0.4$	$y_4 = ?$	

Now, consider Milne's predictor-corrector formulae

$$y_4^{(p)} = y_0 + \frac{4h}{3}(2y_1^1 - y_2^1 + 2y_3^1)$$

$$= 1 + \frac{4(.1)}{3}[2 \times 1.245 - 1.6043 + 2 \times 2.1235]$$

$$= 1.6844.$$

Hence, $y_4^1 = x_4^2 + y_4^2 = (0.4)^2 + (1.6844)^2 = 2.9972$

$$\therefore \quad y_4^{(c)} = y_2 + \frac{h}{3}(y_2^1 + 4y_3^1 + y_4^1)$$

$$= 1.2507 + \frac{0.1}{3}[1.6043 + 4 \times 2.1235 + 2.9972]$$

$$= 1.6874$$

Again, $y_4^1 = (0.4)^2 + (1.6874)^2 = 3.0067$

$$\therefore \quad y_4^{(c)} = 1.2507 + \frac{0.1}{3}[1.6043 + 4 \times 2.1235 + 3.0067]$$

$$= 1.6875$$

Again, $y_4^1 = (0.4)^2 + (1.6875)^2 = 3.0077$

$$\therefore \quad y_4^{(c)} = 1.2507 + \frac{0.1}{3}[1.6043 + 4 \times 2.1235 + 3.0077]$$

$= 1.6876$

Again, $y_4^1 = (0.4)^2 + (1.6876)^2 = 3.008$

$$\therefore \quad y_4^{(c)} = 1.2507 + \frac{0.1}{3}[1.6043 + 4 \times 2.1235 + 3.008]$$

$= 1.6875733$

Thus, $y(0.4) = 1.6876$

Example 4: The following table gives the solution of $5xy^1 + y^2 - 2 = 0$. Find the value of y at $x = 4.5$ using Milne's predictor-corrector formula use corrector formula twice.

x:	4	4.1	4.2	4.3	4.4
y:	1	1.0049	1.0097	1.0143	1.0187

Solution: By data $5x\dfrac{dy}{dx} + y^2 - 2 = 0$ or $y^1 = \dfrac{2 - y^2}{5x}$.

we prepare the table as follows.

x	y	$y^1 = \dfrac{2-4^2}{5x}$
$x_0 = 4$	$y_0 = 1$	$y_0^1 = 0.05$
$x_1 = 4.1$	$y_1 = 1.0049$	$y_1^1 = 0.0483$
$x_2 = 4.2$	$y_2 = 1.0097$	$y_2^1 = 0.0467$
$x_3 = 4.3$	$y_3 = 1.0143$	$y_3^1 = 0.04517$
$x_4 = 4.4$	$y_4 = 1.0187$	$y_4^1 = 0.04374$
$x_5 = 4.5$	$y_5 = ?$	

Since we have to find $y(x_5)$, x_0, y_0 are not necessary to use.

$$\therefore \quad y_5^{(P)} = y_1 + \frac{4h}{3}(2y_2^1 - y_3^1 + 2y_4^1) \qquad \text{(Here } h = 0.1\text{)}$$

$$= 1.0049 + \frac{4\,(.1)}{3}[2 \times 0.0467 - 0.04517 + 2 \times 0.04374]$$

$= 1.023$

$$\text{Now } y_5^1 = \frac{2 - 4^2 5}{5x_5} = \frac{2-(1.023)^2}{5(4.5)} = 0.04237$$

$$\therefore \quad y_5^{(c)} = y_3 + \frac{h}{3}\left(y_3^1 + 4y_4^1 + y_5^1\right)$$

$$= 1.0143 + \frac{.1}{3}[0.04517 + 4 \times 0.04374 + 0.04237]$$

$$= 1.02305$$

$$\text{Again } y_5^1 = \frac{2 - y_5^2}{5x_2} = \frac{2-(1.02305)^2}{5(4.5)} = 0.0423719$$

$$\therefore \quad y_5^{(c)} = y_3 + \frac{h}{3}\left[y_3^1 + 4y_4^1 + y_5^1\right]$$

$$\therefore \quad y_5^{(c)} = 1.02305$$

Hence $y(4.5) = 1.02305$

Example 5: Apply Milne's method, to find a solution of the differential equation $\frac{dy}{dx} = x - y^2$ in the range $0 \leq x \leq 1$ for the boundary condition $y = 0$ at $x = 0$.

Solution: Here $f(xy) = x - y^2$ $x_0 = 0$, $y_0 = 0$.

To determine the starting values of the Milne's method we use Picard's method.

We have,

$$y = y(0) + \int_0^x f(xy)\, dx$$

$$y_1 = 0 + \int_0^x x dx = \frac{x^2}{2}$$

$$\therefore \quad y_2 = y(0) + \int_0^x f(x_1\, y_1)\, dx$$

$$= 0 + \int_0^x \left(x - \frac{x^4}{4}\right) dx = \frac{x^2}{2} - \frac{x^5}{20}$$

$$y_3 = y(0) + \int_0^x f\left(x_1\, y_2\right) dx$$

$$= 0 + \int_0^x x - \left(\frac{x^2}{2} - \frac{x^5}{20}\right)^2 dx$$

$$= \frac{x^2}{2} - \frac{x^5}{20} + \frac{x^8}{160} - \frac{x^{11}}{4400} \ldots \tag{1}$$

Let $h = 0.2$ and from (1)

$$\therefore \quad x = 0.0 \quad y_0 = 0.000 \quad y_0^1 = 0.0000$$

$$x = 0.2 \quad y_1 = 0.0200 \quad y_1^1 = 0.1996$$

$$x = 0.4 \quad y_2 = 0.0795 \quad y_2^1 = 0.3937$$

$$x = 0.6 \quad y_3 = 0.1762 \quad y_3^1 = 0.5689$$

Using predictor formula

$$y_4^{(P)} = y_0 + \frac{4h}{3}\left(2y_1^1 - y_2^1 + 2y_3^1\right)$$

$$= 0.3049$$

Now,

$$y_4^1 = x_4 - y_4^{(8)} = 0.8 - (0.3049)^2 = 0.7070$$

and corrector formula

$$y_4^{(c)} = y_2 + \frac{h}{3}\left(y_2^1 + 4y_3^1 + y_4^1\right)$$

$$= 0.0795 + \frac{0.2}{3}(0.3937 + 4(0.5689) + 0.7070)$$

$$= 0.3046.$$

Again using corrector formula we get

$$y_u^{(c)} = 0.3046.$$

Now,

$$y_5^{(P)} = y_1 + \frac{4h}{3}\left(2y_2' - y_3' + 2y_4'\right)$$

$$= 0.0200 + \frac{4(.2)}{3}(2 \times 0.3937 - 0.5689 + 2 \times 0.3046)$$

$$= 0.4554$$

Now,

$$y_5^1 = x_5 - y_5^2 = 0.7926$$

$$y_5^{(c)} = y_3 + \frac{h}{3}(y_3 + 4y_4 + y_5)$$

$$= 0.1762 + \frac{0.2}{3}(0.5689 + 4(0.3046) + 0.7926)$$

$$= 0.4555$$

$\therefore \quad x = 1, \quad y(1) = 0.4555.$

1.1.6 Adam's Bashforth Predictor-Corrector Method

Let $\frac{dy}{dx} = f(x,y) \ldots (1)$ be the given differential equation to be solved under the conditions $y(x_0) = y_0$.

$$y_{n+1}^{(p)} = y_n + \frac{h}{24}\left[55y_n^1 - 59y_{n-1}^1 + 37y_{n-2}^1 - 9y_{n-3}^1\right]$$

This is Adam's Bashforth predictor formula.

$$y_{n+1}^{(c)} = y_n + \frac{h}{24}\left[9y_{n+1}^1 + 19y_n^1 - 5y_{n-1}^1 + y_{n-2}^1\right]$$

This is Adam's corrector formula.

If $n = 3$

$$y_4^{(p)} = y_3 + \frac{1}{24}\left[55y_3^1 - 59y_2^1 + 37y_1^1 - 9y_0^1\right]$$

and $$y_4^{(c)} = y_3 + \frac{h}{24}\left[9y_4^1 + 19y_3^1 - 5y_2^1 + y_1^1\right]$$

Example 1: Given $\dfrac{dy}{dx} = x - y^2$ and the data

x:	0.0	0.2	0.4	0.6
y:	0.0	0.02	0.0795	0.1762

Compute y at $x = 0.8$ by applying Adam's Bashforth method.

Solution: Here $f(x,y) = x - y^2$ and $h = 0.2$.

First we prepare the following table

x	y	$y^1 = x - y^2$
$x_0 = 0$	$y_0 = 0$	$y_0^1 = 0$
$x_1 = 0.2$	$y_1 = 0.02$	$y_1^1 = 0.1996$
$x_2 = 0.4$	$y_2 = 0.0795$	$y_2^1 = 0.3937$
$x_3 = 0.6$	$y_3 = 0.1762$	$y_3^1 = 0.5689$
$x_4 = 0.8$	$y_4 = ?$	

Consider Adam's Bashforth predictor formula

$$y_4^{(p)} = y_3 + \frac{h}{24}\left[55y_3^1 - 59y_2^1 + 37y_1^1 - 9y_0^1\right]$$

$$= 0.1762 + \frac{0.2}{24}[55 \times 0.5689 - 59 \times 0.3937 + 37 \times 0.1996 - 9 \times 0]$$

$$= 0.3049$$

Now, $y_4^1 = x_4 - y_4^2 = 0.8 - (0.3049)^2 = 0.707$

$$\therefore \quad y_4^{(c)} = y_3 + \frac{h}{24}\left[9y_4^1 + 19y_3^1 - 5y_2^1 + y_1^1\right]$$

$$= 0.1762 + \frac{0.2}{24}[9 \times 0.3049 + 19 \times 0.5689 - 5 \times 0.3937 + 0.1996]$$

$$= 0.3046.$$

Again, $y_4^1 = x_4 - y_4^2 = 0.8 - (0.3046)^2 = 0.7072$

$$y_4^{(c)} = y_3 + \frac{h}{24}\left[9y_4^1 + 19y_3^1 - 5y_2^1 + y_1^1\right]$$

$$= 0.1762 + \frac{0.2}{24}[9 \times 0.7072 + 19 \times 0.5689 - 5 \times 0.3937 + 0.1996]$$

$$= 0.3046$$

$$\therefore \quad y(0.8) = 0.3046$$

Example 2: Solve the differential equation $\frac{dy}{dx} = x^2 - y$ with the condition that $y_0 = 1,\ y_1 = 0.9052,\ y_2 = 0.8213,\ y_3 = 0.7492$. Find the value of y_4 and y_5 corresponding to $x_4 = 0.4$ and $x_5 = 0.5$. Apply Adam's Bashforth predictor-corrector formula.

Solution: Given $\frac{dy}{dx} = x^2 - y.$

First we prepare a table as follows.

x	y	$y^1 = x^2 - y$
$x_0 = 0$	$y_0 = 1$	$y_0^1 = -1$
$x_1 = 0.1$	$y_1 = 0.9052$	$y_1^1 = -0.8952$
$x_2 = 0.2$	$y_2 = 0.8213$	$y_2^1 = -0.7813$
$x_3 = 0.3$	$y_3 = 0.7492$	$y_3^1 = -0.6592$
$x_4 = 0.4$	$y_4 = ?$	
$x_5 = 0.5$	$y_5 = ?$	

First we find y_4 at $x = 0.4$

By Adam's Bashforth predictor formula we have

$$y_u^{(P)} = y_3 + \frac{h}{24}\left[55y_3^1 - 59y_2^1 + 374_1^1 - 97y_0^1\right]$$

$$= 0.6897$$

$$\text{Now,} \quad y_4^1 = x_4^2 - y_4 = (0.4)^2 - 0.6897 = -0.5297$$

$$\therefore \quad y_4^{(c)} = y_3 + \frac{h}{24}\left[9y_4^1 + 19y_3^1 - 5y_2^1 + y_1^1\right]$$

$$= 0.6897$$

Thus, when $x = 0.4$, $y = 0.6897$.

Now, to compute y_5 we shall consider equivalent form of $y^{(p)}$ and $y^{(c)}$.

$$y_5^{(p)} = y_4 + \frac{h}{24}\left[55y_4^1 - 59y_3^1 + 37y_2^1 - 9y_1^1\right]$$

$$= 0.6435$$

$$y_5^1 = x_5^2 - y_5 = (0.5)^2 - 0.6435 = -0.3935$$

$$\therefore \quad y_5^{(c)} = y_4 + \frac{h}{24}\left[9y_5^1 + 19y_4^1 - 5y_3^1 + y_2^1\right]$$

$$= 0.6435$$

Thus $y_5 = y(0.5) = 0.6435$

QUESTION BANK

1. Using Picard's method obtain the solution for the differential equation $\frac{dy}{dx} = x^2 + y^2$ for $x = 0.4$ given that $y = 0$ when $x = 0$.
   Ans: $y(0.4) = 0.0214$
2. Obtain Picard's second approximate solution of the initial value problem $\frac{dy}{dx} = \frac{x^2}{y^2 + 1}$ given $y(0) = 0$.
   Ans: $y = \frac{1}{3}x^3 - \frac{1}{81}x^9$
3. Find approximate value of y when $x = 0.1$, if $\frac{dy}{dx} = x - y^2$ and $y = 1$ at $x = 0$ using Picard's method.
   Ans: $y(0.1) = 0.9138$
4. Using Picards method solve $y^1 = y^2 + x$ given that $y(0) = 1$ for computing $y(0.1)$ and $y(0.2)$.
   Ans: $y(0.1) = 1.1164$, $y(0.2) = 1.2725$
5. Solve by Taylor's series method $\frac{dy}{dx} = x + y$ start from $x = 1$, $y = 0$ and carry to $x = 1.2$ with $h = 0.1$ compare the final value with the exact value.
   Ans: $y(1.1) = 0.1103384$, $y(1.2) = 0.232$

   Exact value $y(1.2) = 0.242$

6. Solve $\frac{dy}{dx} = y^2 + 1$ assuming $y = 0$ at $x = 0$ in the range $0 \le x \le 1$ taking step size 0.2 by Taylor's series method.
Ans: $y(0.2) = 0.203$, $y(0.4) = 0.422$, $y(0.6) = 0.683$, $y(0.8) = 1.027$, $y(1) = 1.546$

7. Solve $y^1 = y^2 + x$, $y(0) = 1$ using Taylor's series method to compute $y(0.1)$ and $y(0.2)$.
Ans: $y(0.1) = 1.1164$, $y(0.2) = 1.2725$

8. Find approximate value of y when $x = 0.1$ if $\frac{dy}{dx} = x - y^2$ and $y = 1$ at $x = 0$ by Taylor's series method.
Ans: $y(0.1) = 0.9188$

9. Given $\frac{dy}{dx} = \frac{y-x}{y+x}$ with boundary conditions $y = 1$ when $x = 0$, find approximately y for $x = 0.1$ by Euler's modified method (5 steps).
Ans: $y(0.1) = 1.0928$

10. Solve the differential equation $\frac{dy}{dx} = -xy^2$, $y = 2$ at $x = 0$ by Euler's modified method and obtain y at $x = 0.2$ in two stages of 0.1 each.
Ans: $y(0.2) = 1.9227$

11. Given that $\frac{dy}{dx} = 2 + \sqrt{xy}$ and $y = 1$ when $x = 1$. Find approximate value of y at $x = 2$ in steps of 0.2, using Euler's modified method.
Ans: $y(2) = 5.051$

12. Using Euler's method solve for y at $x = 0.1$ from $\frac{dy}{dx} = x + y + xy$, $y(0) = 1$ taking step size $h = 0.025$.
Ans: $y(0.1) = 1.1448$

13. Using Runge-Kutta method of order four compute $y(0.2)$ and $y(0.4)$ from $10\frac{dy}{dx} = x^2 + y^2$, $y(0) = 1$ and taking $h = 0.1$.
Ans: $y(0.2) = 1.0207$, $y(0.4) = 1.0438$

14. Using Runge-Kutta method of order four find $y(0.2)$ for the equation $\frac{dy}{dx} = \frac{y-x}{y+x}$, $y(0) = 1$ and taking $h = 0.2$.
Ans: $y(0.2) = 1.1678$

15. Using Runge-Kutta method of order four find y at $x = 0.8$ given that $y = 0.41$ when $x = 0.4$ and $\frac{dy}{dx} = \sqrt{x+y}$.

    Ans: $y(0.8) = 0.8489$

16. Solve numerically using Runge-Kutta fourth order formula the differential equation $\frac{dy}{dx} = x^2 + y^2$ with given conditions $x = 1$, $y = 1.5$ in the interval $(1, 1.2)$ with $h = 0.1$.

    Ans: $y(1.1) = 1.8955$, $y(1.2) = 2.5043$

17. Given $\frac{dy}{dx} = \frac{(1-x^2)y^2}{2}$ and $y(0) = 1$, $y(0.1) = 1.06$, $y(0.2) = 1.12$, $y(0.3) = 1.21$, evaluate $y(0.4)$ by Milne's predictor and corrector method.

    Ans: $y(0.4) = 1.2797$

18. Use Milne's method to find $y(0.3)$ from $y^1 = x^2 + y^2, y(0) = 1$. Find the initial values $y(-0.1), y(0.1)$ and $y(0.2)$ from the Taylor's series method.

    Ans: $y(0.3) = 1.4392$

19. Using Adam's-Bashforth formulae, determine $y(0.4)$ given the differential equation $\frac{dy}{dx} = \frac{1}{2}xy$ and the data

x:	0	0.1	0.2	0.3
y:	1	1.0025	1.0101	1.0228

    Ans: $y(0.4) = 1.0408$

20. Use Adam's method to find $y(0.4)$ given that $\frac{dy}{dx} = x + y$ and $y(0) = 1, y(0.1) = 1.1034, y(0.2) = 1.2428, y(0.3) = 1.3997$. Apply corrector formula twice.

    Ans: $y(0.4) = 1.5836$

Numerical Methods – II

UNIT 2

2.1 NUMERICAL SOLUTION OF SIMULTANEOUS FIRST ORDER ORDINARY DIFFERENTIAL EQUATIONS

2.1.1 Runge-Kutta Method of Fourth Order

1. Let $\frac{dy}{dx} = f(x,y,z)$, $\frac{dz}{dx} = g(x,y,z)$ be the simultaneous first order ordinary differential equations with the initial values $y(x_0) = y_0$, $z(x_0) = z_0$. The modified form of expression for computation are as follows.

$$k_1 = hf(x_0, y_0, z_0); \qquad l_1 = hg(x_0, y_0, z_0)$$

$$k_2 = hf\left(x_0 + \frac{h}{2}, y_0 + \frac{k_1}{2}, z_0 + \frac{l_1}{2}\right); \qquad l_2 = hg\left(x_0 + \frac{h}{2}, y_0 + \frac{k_1}{2}, z_0 + \frac{l_1}{2}\right)$$

$$k_3 = hf\left(x_0 + \frac{h}{2}, y_0 + \frac{k_2}{2}, z_0 + \frac{l_2}{2}\right); \quad l_3 = hg\left(x_0 + \frac{h}{2}, y_0 + \frac{k_2}{2}, z_0 + \frac{l_2}{2}\right)$$

$$k_4 = hf\left(x_0 + h, y_0 + k_3, z_0 + l_3\right); \qquad l_4 = hg\left(x_0 + h, y_0 + k_3, z_0 + l_3\right)$$

Then according to second order Runge-Kutta method

$$y(x_0 + h) = y_0 + k_2$$

$$z(x_0 + h) = z_0 + l_2$$

According to fourth order Runge-Kutta method

$$y(x_0 + h) = y_0 + \frac{1}{6}\left(k_1 + 2k_2 + 2k_3 + k_4\right) \quad \text{and}$$

$$z(x_0 + h) = z_0 + \frac{1}{6}\left(l_1 + 2l_2 + 2l_3 + l_4\right)$$

Example 1: Use fourth order Runge-Kutta method to solve the system of equations

$$\frac{dx}{dt} = y - t, \frac{dy}{dt} = x + t \quad \text{given } x = 1,\ y = 1 \text{ at } t = 0$$

Compute $x(0.1)$ and $y(0.1)$.

Solution: The simultaneous equations are

$$\frac{dx}{dt} = y - t, \frac{dy}{dt} = x + t \quad \text{with } x = 1,\ y = 1 \text{ at } t = 0$$

Let us take $h = 0.1$. Since we need to compute $x(0.1)$ and $y(0.1)$.

First we shall compute the following.

$$f(t, x, y) = y - t, \quad y(t, x, y) = x + t$$

$$k_1 = hf(t_0, x_0, y_0) = 0.1(1 - 0) = 0.1$$

$$l_1 = hg(t_0, x_0, y_0) = 0.1(1 + 0) = 0.1$$

$$k_2 = hf\left(t_0 + \frac{h}{2}, x_0 + \frac{k_1}{2}, y_0 + \frac{l_1}{2}\right)$$

$$= 0.1\,[1.05 - 0.05] = 0.1$$

$$l_2 = hg\left(t_0 + \frac{h}{2}, x_0 + \frac{k_1}{2}, y_0 + \frac{l_1}{2}\right)$$

$$= 0.1\,[1.05 + 0.05] = 0.11$$

$$k_3 = hf\left(t_0 + \frac{h}{2}, x_0 + \frac{k_2}{2}, y_0 + \frac{l_2}{2}\right)$$

$$= 0.1\,[1.055 - 0.05] = 0.1005$$

$$l_3 = hg\left(t_0 + \frac{h}{2}, x_0 + \frac{k_2}{2}, y_0 + \frac{l_2}{2}\right)$$

$$= 0.1\,[1.05 + 0.05] = 0.11$$

$$k_4 = hf\left(t_0 + h, x_0 + k_3, y_0 + l_3\right)$$

$$= .1\,[1.11 - 0.1] = 0.101$$

$$l_4 = hg\left(t_0 + h, x_0 + k_3, y_0 + l_3\right)$$

$$= 0.1\,[1.1005 + 0.1] = 0.12005$$

$$\therefore \quad x(t_0 + h) = x_0 + \frac{1}{6}\left[k_1 + 2k_2 + 2k_3 + k_4\right]$$

$$= 1 + \frac{1}{6}[0.1 + 2 \times 0.1 + 2 \times 0.1005 + 0.101]$$

$$= 1.100333$$

and

$$y(t_0 + h) = y_0 + \frac{1}{6}\left[l_1 + 2l_2 + 2l_3 + l_4\right]$$

$$= 1 + \frac{1}{6}[0.1 + 2(0.11) + 2(0.11) + 0.12005]$$

$$= 1.1100083$$

Thus,

$$x(0.1) = 1.100333 \simeq 1.1$$

$$y(0.1) = 1.1100083 \simeq 1.11.$$

Example 2: Use fourth order Runge-Kutta method to solve the system of equations

$$\frac{dy}{dx} = 1 + xz; \ \frac{dz}{dx} = -xy \quad \text{given } y(0) = 0,\ z(0) = 1 \text{ at } x = 0.3.$$

Solution: Given simultaneous equations are

$$\frac{dy}{dx} = 1 + xz, \frac{dz}{dx} = -xy \quad \text{and } y(0) = 0,\ z(0) = 1$$

or, $f(x,y,z) = 1 + xz, \quad g(x,y,z) = -xy$

Here $x_0 = 0,\ y_0 = 0,\ z_0 = 1$ and given $h = 0.3$.

Now, we shall compute the following.

$$k_1 = hf(x_0, y_0, z_0) = 0.3\,[1 + 1 \times 0] = 0.3$$

$$l_1 = hg(x_0, y_0, z_0) = 0.3\,[-0 \times 0] = 0$$

$$k_2 = hf\left(x_0 + \frac{h}{2}, y_0 + \frac{k_1}{2}, z_0 + \frac{l_1}{2}\right)$$

$$= 0.3\,[1 + (1)(0.15)] = 0.345$$

$$l_2 = hg\left(x_0 + \frac{h}{2}, y_0 + \frac{k_1}{2}, z_0 + \frac{l_1}{2}\right)$$

$$= 0.3\,[-(0.15)(0.15)] = -0.00675$$

$$k_3 = hf\left(x_0 + \frac{h}{2}, y_0 + \frac{k_2}{2}, z_0 + \frac{l_2}{2}\right)$$

$$= 0.3\,[(1 + 0.996625)(0.15)] = 0.34485$$

$$l_3 = hf\left(x_0 + \frac{h}{2}, y_0 + \frac{k_2}{2}, z_0 + \frac{l_2}{2}\right)$$

$$= 0.3\,[-(0.15)(0.1725)] = -0.0077625$$

$$k_4 = hf\,(x_0 + h, y_0 + k_3, z_0 + l_3)$$

$$= 0.3\,[1 + (0.99224)(0.3)] = 0.3893$$

$$l_4 = hf\,(x_0 + h, y_0 + k_3, z_0 + l_3)$$

$$= 0.3\,[-0.3(.34485)] = -0.03104$$

$$\therefore \quad y(x_0 + h) = y_0 + \frac{1}{6}\,[k_1 + 2k_2 + 2k_3 + k_4]$$

$$\therefore \quad y(0.3) = 0 + \frac{1}{6}\,[.3 + 2(.345) + 2(0.34485) + 0.3893]$$

$$= 0.34483$$

and

$$z(x_0 + h) = z_0 + \frac{1}{6}[l_1 + 2l_2 + 2l_3 + l_4]$$

$$= 1 + \frac{1}{6}[0 + 2(-0.00675) + 2(-0.0077625) + (-0.03104)]$$

$$\therefore \quad z(0.3) = 0.98999.$$

2.1.2 Picard's Method

Let the simultaneous differential equations be

$$\frac{dy}{dx} = f(x,y,z) \text{ and } \frac{dz}{dx} = g(x,y,z)$$

with initial conditions $y(x_0) = y_0$ and $z(x_0) = z_0$. Then

$$y_1 = y_0 + \int f(x,y_0,z_0)\,dx, \qquad z_1 = z_0 + \int g(x,y_0,z_0)\,dx$$

$$y_2 = y_0 + \int f(x,y_1,z_1)\,dx, \qquad z_2 = z_0 + \int g(x,y_1,z_1)\,dx$$

$$y_3 = y_0 + \int f(x,y_2,z_2)\,dx, \qquad z_3 = z_0 + \int g(x,y_2,z_2)\,dx$$

and so on.

Continuing this process, a sequence of functions of x, i.e., $y_1, y_2, y_3 \ldots$ and $z_1, z_2, z_3 \ldots$ are obtained each giving a better approximations of the desired solution.

Example 1: Using Picard's method find approximate values of y and z corresponding to $x = 0.1$ given that $y(0) = 2$, $z(0) = 1$ and $\frac{dy}{dx} = x + z$, $\frac{dz}{dx} = x - y^2$.

Solution: Given equations are

$$\frac{dy}{dx} = f(x,y,z) = x + z \quad \text{and} \quad \frac{dz}{dx} = g(x,y,z) = x - y^2$$

$$\therefore \quad y = y_0 + \int_{x_0}^{x} f(x,y,z)\,dx \text{ and } z = z_0 + \int_{x_0}^{x} g(x,y,z)\,dx$$

First approximation

$$y_1 = y_0 + \int_{x_0}^{x} f(x, y_0, z_0)\, dx = 2 + \int_{0}^{x} (x+1)\, dx = 2 + x + \frac{x^2}{2}$$

$$z_1 = z_0 + \int_{x_0}^{x} g(x, y_0, z_0)\, dx = 1 + \int_{0}^{x} (x-4)\, dx = 1 - 4x + \frac{1}{2}x^2$$

Second approximation

$$y_2 = y_0 + \int_{x_0}^{x} f(x, y_1, z_1) dx = 2 + \int_{0}^{x} \left(x + 1 - 4x + \frac{1}{2}x^2 \right) dx$$

$$= 2 + x - \frac{3}{2}x^2 + \frac{x^3}{6}$$

$$z_2 = z_0 + \int_{x_0}^{x} g(x, y_1, z_1) dx = 1 + \int_{0}^{x} \left[x - \left(2 + x + \frac{1}{2}x^2 \right)^2 \right] dx$$

$$= 1 + 4x + \frac{3}{2}x^2 - x^3 - \frac{x^4}{4} - \frac{x^5}{20}$$

Third approximation

$$y_3 = y_0 + \int_{x_0}^{x} f(x, y_2, z_2) dx$$

$$= 2 + x - \frac{3}{2}x^2 - \frac{1}{2}x^3 - \frac{1}{4}x^4 - \frac{1}{20}x^5 - \frac{1}{120}x^6$$

$$z_3 = z_0 + \int_{0}^{x} g(x, y_2, z_2) dx$$

$$= 1 - 4x - \frac{3}{2}x^2 - \frac{5}{3}x^3 + \frac{7}{12}x^4 - \frac{31}{60}x^5 + \frac{1}{12}x^6 - \frac{1}{252}x^7$$

and so on.

when $x = 0.1$

$$y_1 = 2.105, \quad y_2 = 2.08517 \quad y_3 = 2.08447$$

$$z_1 = 0.605, \quad z_2 = 0.58397 \quad z_3 = 0.58672$$

Thus, $y(0.1) = 2.0845$, $z(0.1) = 0.5867$ correct to four decimal places.

2.1.3 Taylor's Series Method

Let the simultaneous differential equations be

$$\frac{dy}{dx} = f(x,y,z) \quad (1)$$

$$\text{and} \quad \frac{dz}{dx} = g(x,y,z) \quad (2)$$

with initial conditions $y(x_0) = y_0$ and $z(x_0) = z_0$.

If h be the step-size, $y_1 = y(x_0 + h)$ and $z_1 = z(x_0 + h)$

Then Taylor's algorithm for (1) and (2) gives

$$y_1 = y_0 + hy_0' + \frac{h^2}{2!}y_0'' + \frac{h^3}{3!}y_0''' \dots \quad (3)$$

$$z_1 = z_0 + hz_0' + \frac{h^2}{2!}z_0'' + \frac{h^3}{3!}z_0''' \dots \quad (4)$$

Differentiating (1) and (2) successively we get y'', z'', etc.

So the values $y_0', y_0'', y_0''' \dots$ and $z_0', z_0'', z_0''' \dots$ are known. Substituting these in (3) and (4), we obtain y_1, z_1 for the next step.

Similarly,

$$y_2 = y_1 + hy_1' + \frac{h^2}{2!}y_1'' + \frac{h^3}{3!}y_1''' \dots \quad (5)$$

$$z_2 = z_1 + hz_1' + \frac{h^2}{2!}z_1'' + \frac{h^3}{3!}z_1''' \dots \quad (6)$$

Since y_1 and z_1 are known, we calculate $y_1', y_1'' \dots$ and $z_1', z_1'' \dots$ substituting these in (5) and (6) we get y_2 and z_2.

Proceeding further, we can calculate the other values of y and z step by step.

Example 1: Given $\frac{dy}{dx} = z$ and $\frac{dz}{dx} = -xz - y$ with $y(0) = 1$, $z(0) = 0$, obtain y and z for $x = 0.1, 0.2, 0.3$ by Taylor's series method.

Hint:

We have

$$y' = z \quad \text{and} \quad z' = -xz - y$$

We use Taylor's series method to find y and z.

2.2 NUMERICAL SOLUTION OF SECOND ORDER ORDINARY DIFFERENTIAL EQUATIONS

2.2.1 Picard's Method

Let the second order differential equation be

$$\frac{d^2y}{dx^2} = f\left(x, y, \frac{dy}{dx}\right) \tag{1}$$

with $y(x_0) = y_0$ and $y'(x_0) = y'_0$.

Now, let $\frac{dy}{dx} = z$ then $\frac{d^2y}{dx^2} = \frac{dz}{dx}$.

Substituting in (1)

$$\frac{dz}{dx} = f(x, y, z) \text{ with } y(x_0) = y_0 \text{ and } z(x_0) = z_0$$

The problem reduces to solving the simultaneous equations

$$\frac{dy}{dx} = z = f_1(x, y, z)$$

$$\frac{dz}{dx} = f_2(x, y, z) \text{ subject to } y(x_0) = y_0 \text{ and } z(x_0) = z_0.$$

By Picard's method we have

$$y_1 = y_0 + \int_{x_0}^{x} f_1(x, y_0, z_0)\, dx, \quad z_1 = z_0 + \int_{x_0}^{x} f_2(x, y_0, z_0)\, dx$$

$$y_2 = y_0 + \int_{x_0}^{x} f_1(x, y_1, z_1)\, dx, \quad z_2 = z_0 + \int_{x_0}^{x} f_2(x, y_1, z_1)\, dx$$

$$y_3 = y_0 + \int_{x_0}^{x} f_1(x, y_2, z_2)\, dx, \quad z_3 = z_0 + \int_{x_0}^{x} f_2(x, y_2, z_2)\, dx \text{ and so on.}$$

Continuing this process, a sequence of functions x, i.e., $y_1, y_2, y_3 \ldots$; $z_1, z_2, z_3 \ldots$ is obtained each giving a better approximation of the desired solution than the preceding one.

Example 1: Use Picard's method to approximate y when $x = 0.1$ given that $\frac{d^2y}{dx^2} + 2x\frac{dy}{dx} + y = 0$ and $y = 0.5$, $\frac{dy}{dx} = 0.1$ when $x = 0$.

Solution: Given equation of second order is

$$\frac{d^2y}{dx^2} + 2x\frac{dy}{dx} + y = 0$$

or
$$\frac{d^2y}{dx^2} = -2x\frac{dy}{dx} - y \tag{1}$$

with $y_0 = 0.5$, $y'_0 = 0.1$ when $x = 0$.

Let $\frac{dy}{dx} = z$ then $\frac{d^2y}{dx^2} = \frac{dz}{dx}$

$\therefore \quad \frac{dy}{dx} = z = f_1(x, y, z)$ and $\frac{dz}{dx} = -2xz - y = f_2(x, y, z)$ (by (1)) with $y_0 = 0.5$, $z_0 = 0.1$ when $x = 0$.

By Picard's method, we have

$$y_1 = y_0 + \int_{x_0}^{x} f_1(x, y_0, z_0)\, dx, \qquad z_1 = z_0 + \int_{x_0}^{x} f_2(x, y_0, z_0)\, dx$$

$$= 0.5 + \int_0^x 0.1\, dx \qquad\qquad = 0.1 + \int_0^x (-.2x - .5)\, dx$$

$$= 0.5 + 0.1x \qquad\qquad = 0.1 - \frac{.2x^2}{2} - .5x$$

$$y_2 = y_0 + \int_{x_0}^{x} f_1(x, y_1, z_1)\, dx \qquad z_2 = z_0 + \int_{x_0}^{x} f_2(x, y_1, z_1)\, dx$$

$$= 0.5 + \int_0^x \left(.1 - 0.2\frac{x^2}{2} - 0.5x \right) dx, \quad z_2 = 0.1 + \int_0^x -2x\left(0.1 - \frac{.2}{2}x^2 - .5x\right)$$

$$- (0.5 + 0.1x)\, dx$$

$$= 0.5 + 0.1x - 0.2\frac{x^3}{6} - 0.5\frac{x^2}{2}, \qquad = 0.1 - .2\frac{x^2}{2} + 0.2\frac{x^4}{4} + 1\frac{x^3}{3}$$

$$- 0.5x - 0.1\frac{x^2}{2}$$

and so on.

when $x = 0.1$

$$y_1 = 0.51 \qquad y_2 = 0.5074$$

$$z_1 = y_1' = 0.049 \qquad y_2' = z_2 = 0.0488$$

$$\therefore \quad y(0.1) = 0.5074$$

$$y'(0.1) = 0.0488$$

Example 2: Using Picard's method, obtain the second approximation to the solution of

$$\frac{d^2y}{dx^2} = x^3\frac{dy}{dx} + x^3 y \quad \text{so that } y(0) = 1, \quad y'(0) = \frac{1}{2}.$$

Solution: Given the second order differential equation is

$$\frac{d^2y}{dx^2} = x^3\frac{dy}{dx} + x^3 y \quad \text{with} \quad y(0) = 1, \quad y'(0) = \frac{1}{2}$$

Let $\dfrac{dy}{dx} = z$ so that $\dfrac{d^2y}{dx^2} = \dfrac{dz}{dx}$

$$\therefore \quad \frac{dy}{dx} = z \qquad \text{and} \quad \frac{dz}{dx} = x^3 z + x^3\, y \quad \text{with } y(0) = 1, \quad z(0) = \frac{1}{2}.$$

$$= f(x,y,z) \qquad = f_1(x,y,z)$$

By Picard's methods,

First approximation

$$y_1 = y_0 + \int_{x_0}^{x} f(x\, y_0\, z_0)dx \qquad z_1 = z_0 + \int_{x_0}^{x} f_2(x,\, y_0\, z_0)\, dx$$

$$= 1 + \int_0^x \frac{1}{2}dx \qquad = \frac{1}{2} + \int_0^x \left(\frac{1}{2}x^3 + 1x^3\right) dx = \frac{1}{2} + \int \frac{3}{2}x^3 dx$$

$$= 1 + \frac{x}{2} \qquad = \frac{1}{2} + \frac{3x^4}{8}$$

Second approximation

$$y_2 = y_0 + \int_{x_0}^{x} f(x,\, y_1,\, z_1)\, dx \qquad z_2 = z_0 + \int_{x_0}^{x} f(x, y_1, z_1)\, dx$$

$$= 1 + \int_0^x \left(\frac{1}{2} + \frac{3x^4}{8}\right) dx \qquad = \frac{1}{2} + \int_0^x \left(x^3\left(\frac{1}{2} + \frac{3x^4}{8}\right) + x^3\left(1 + \frac{x}{2}\right)\right) dx$$

$$= 1 + \frac{x}{2} + \frac{3x^5}{x_0} \qquad = \frac{1}{2} + \int_0^x \left(\frac{x^3}{2} + \frac{3}{8}x^7 + x^3 + \frac{x^4}{2}\right) dx$$

$$= \frac{1}{2} + \frac{x^4}{8} + \frac{3x^8}{64} + \frac{x^4}{4} + \frac{x^5}{10}$$

$$= \frac{1}{2} + \frac{3x^4}{8} + \frac{x^5}{10} + \frac{3x^8}{64}.$$

Example 3: Using Picard's method find the solution of the equation

$$y'' + xy' + y = 0, \text{ with } y(0) = 1 \quad y'(0) = 0 \quad \text{for } x = 0.1 \text{ and } 0.3.$$

Solution: Given equation is

$$y'' + xy' + y = 0 \quad \text{with } y(0) = 1, \quad y'(0) = 0.$$

Let $\frac{dy}{dx} = y' = z$ so that $\frac{d^2y}{dx^2} = y'' = \frac{dz}{dx}$

$$\therefore \quad \frac{dy}{dx} = z \quad \text{and} \quad \frac{dz}{dx} = -xz + y \quad \text{with } y(0) = 1,\ z(0) = 0$$

$$= f(x,y,z), \quad = f_1(x,y,z).$$

By Picard's method

First approximation

$$y_1 = y_0 + \int_{x_0}^{x} z_0 \, dx \qquad z_1 = z_0 + \int_{x_0}^{x} (-xz_0 - y_0) \, dx$$

$$= 1 + 0 \qquad = 0 + \int_{0}^{x} -1 \, dx$$

$$= 1 \qquad = -x$$

Second approximation

$$y_2 = y_0 + \int_{x_0}^{x} z_1 \, dx \qquad z_2 = z_0 + \int_{x_0}^{x} (-xz_1 - y_1) \, d_x$$

$$= 1 + \int_{0}^{x} -x \, dx \qquad = 0 + \int_{0}^{x} \left(x^2 - 1\right) dx$$

$$= 1 - \frac{x^2}{2} \qquad = \frac{x^3}{3} - x$$

Third approximation

$$y_3 = y_0 + \int_{x_0}^{x} z_2 \, dx \qquad z_3 = z_0 + \int_{x_0}^{x} (-xz_2 - y_2) \, dx$$

$$= 1 + \int_{0}^{x} \left(\frac{x^3}{3} - x\right) dx \qquad = 0 + \int_{0}^{x} \left[-x\left(\frac{x^3}{3} - x\right) - \left(1 - \frac{x^2}{2}\right)\right] dx$$

$$= 1 + \frac{x^4}{12} - \frac{x^2}{2} \qquad -\int_{0}^{x} \left(\frac{x^4}{3} - x^2 + 1 - \frac{x^2}{2}\right) dx$$

$$= 1 - \frac{x^2}{2} + \frac{x^4}{12} \qquad = -\left(\frac{x^5}{15} - \frac{x^3}{3} + x - \frac{x^3}{6}\right)$$

$$= -x + \frac{x^3}{3} - \frac{x^5}{15}$$

when $x = 0.1$

$$y_1(0.1) = 1 \qquad y_2(0.1) = 0.955, \quad y_3(0.1) = 0.99501$$

$$z_1(0.1) = -0.1 \qquad z_2(0.1) = -0.099, \quad z_3(0.1) = -0.0995$$

$$\therefore \quad y(0.1) = 0.99501$$

$$z(0.1) = y'(0.1) = -0.0995$$

when $x = 0.3$

$$y_1(0.3) = 1 \qquad y_2(0.3) = 0.955 \quad y_3(0.3) = 0.9556$$

$$z_1(0.3) = -0.3 \qquad z_2(0.3) = -0.291, \quad z_3(0.3) = -0.2911$$

$$\therefore \quad y(0.3) = 0.9556$$

$$z(0.3) = y'(0.3) = -0.2911.$$

2.2.2 Runge-Kutta Method

Let the second order differential equation be

$$\frac{d^2y}{dx^2} = f\left(x,\, y,\, \frac{dy}{dx}\right) \ldots (1) \text{ with } y(x_0) = y_0,\, y'(x_0) = y_0'.$$

Now, let $\frac{dy}{dx} = z$ so that $\frac{d^2y}{dx^2} = \frac{dz}{dx}$

$$\therefore \quad \frac{dy}{dx} = z \quad \text{and} \quad \frac{dz}{dx} = f(x,y,z)$$

$$= f_1(x,y,z), \qquad = f_2(x,y,z) \text{ with } y(x_0) = y_0,\, z(x_0) = z_0.$$

The problem reduces to solving the simultaneous equations

$$\frac{dy}{dx} = z = f_1(x,y,z) \text{ and } \frac{dz}{dx} = f_2(x,y,z) \text{ with } y(x_0) = y_0 \text{ and } z(x_0) = z_0.$$

Starting at (x_0, y_0, z_0) and taking the step-sizes for x, y, z to be h, k, l respectively, the Runge-Kutta method gives

$$K_1 = hf_1(x_0, y_0, z_0) \qquad l_1 = h\,f_2(x_0, y_0, z_0)$$

$$K_2 = hf_1\left(x_0 + \frac{1}{2}h,\ y_0 + \frac{k_1}{2},\ z_0 + \frac{l_1}{2}\right) \quad l_2 = hf_2\left(x_0 + \frac{h}{2},\ y_0 + \frac{k_1}{2},\ z_0 + \frac{l_1}{2}\right)$$

$$k_3 = hf_1\left(x_0 + \frac{h}{2},\ y_0 + \frac{k_2}{2},\ z_0 + \frac{l_2}{2}\right) \quad l_3 = hf_2\left(x_0 + \frac{h}{2},\ y_0 + \frac{k_2}{2},\ z_0 + \frac{l_2}{2}\right)$$

$$k_4 = hf_1(x_0 + h,\ y_0 + k_3,\ z_0 + l_3) \qquad l_4 = hf_2(x_0 + h,\ y_0 + h_3,\ z_0 + l_3)$$

$$\therefore \quad y_1 = y_0 + \frac{1}{6}(k_1 + 2k_2 + 2k_3 + k_4) \quad z_1 = z_0 + \frac{h}{6}(l_1 + 2l_2 + 2l_3 + l_4)$$

Example 1: Using Runge-Kutta method solve

$y'' = xy'^2 - y^2$ for $x = 0.2$ correct to 4 decimal places initial conditions are $x = 0,\ y = 1,\ y' = 0$.

Solution: Given equation of second order is

$$y'' = xy'^2 - y^2 \quad \text{with } x = 0,\ y = 1,\ y' = 0$$

Let $\dfrac{dy}{dx} = z = f_1(x, y, z)$, then

$$\frac{dz}{dx} = xz^2 - y^2 = f_2(x, y, z) \quad \text{with } x_0 = 0,\ y_0 = 1,\ z_0 = 0$$

we compute k_1, k_2, k_3, k_4 for $f_1(x, y, z)$ and l_1, l_2, l_3, l_4 for $f_2(x, y, z)$.

$\therefore$ by Runge-Kutta formulae, we have

$$k_1 = hf_1(x_0, y_0, z_0) \qquad l_1 = hf_2(x_0, y_0, z_0)$$

$$= 0.2(0) = 0 \qquad = 0.2(-1) = -0.2$$

$$k_2 = hf_1\left(x_0 + \frac{h}{2},\ y_0 + \frac{k_1}{2},\ z_0 + \frac{l_1}{2}\right) \quad l_2 = hf_2\left(x_0 + \frac{h}{2},\ y_0 + \frac{k_1}{2},\ z_0 + \frac{l_1}{2}\right)$$

$$= 0.2(-0.1) = -0.02 \qquad = 0.2(-0.999) = -0.1998$$

$$k_3 = hf_1\left(x_0 + \frac{h}{2},\ y_0 + \frac{k_2}{2},\ z_0 + \frac{l_2}{2}\right) \quad l_3 = hf_2\left(x_0 + \frac{h}{2},\ y_0 + \frac{k_2}{2},\ z_0 + \frac{l_2}{2}\right)$$

$$= 0.2(-0.0999) = -0.02 \qquad = 0.2(-0.9791) = -0.1958$$

$$k_4 = hf_1(x_0 + h,\ y_0 + k_3,\ z_0 + l_3) \qquad l_4 = hf_2(x_0 + h,\ y_0 + k_3,\ z_0 + l_3)$$

$$= 0.2(-0.1958) = -0.0392 \qquad = .2(0.9527) = -0.1905$$

$$\therefore \quad k = \frac{1}{6}(k_1 + 2k_2 + 2h_3 + k_4) \qquad l = \frac{1}{6}(l_1 + 2l_2 + 2l_3 + l_4)$$

$$= -0.0199 \qquad = -0.1970$$

Hence, at $x = 0.2$

$$y = y_0 + k = 1 - 0.0199$$

$$= 0.9801$$

$$z = y' = z_0 + l = 0 - 0.1970$$

$$= -0.1970.$$

Example 2: Using Runge-Kutta method of order four, solve

$$\frac{d^2y}{dx^2} - x\frac{dy}{dx} - y = 0;\ y(0) = 1,\ y'(0) = 0 \text{ to find } y(0.2) \text{ and } y'(0.2).$$

Ans: $y(0.2) = 0.9802,\ y'(0.2) = -0.196.$

Example 3: The angular displacement of θ of a simple pendulum is given by the equation

$$\frac{d^2\theta}{dt^2} + \frac{q}{l}\sin\theta = 0$$

where $l = 98$ cm and $g = 980\ \text{cm/sec}^2$. If $\theta = 0$ and $\frac{d\theta}{dt} = 4.472$ at $t = 0$, use Runge-Kutta method to find θ and $\frac{d\theta}{dt}$ when $t = 0.2$ sec.

Ans: $\theta(0.2) = 0.8367 \quad \frac{d\theta}{dt}(0.2) = 3.6545.$

2.2.3 Milne's Method

Let the second order differential equation be

$$\frac{d^2y}{dx^2} = f\left(x, y, \frac{dy}{dx}\right) \ldots (1) \text{ with } y(x_0) = y_0 \text{ and } y'(x_0) = y'_0.$$

Let $\frac{dy}{dx} = z$ then $\frac{d^2y}{dx^2} = \frac{dz}{dx}$.

Substituting in (1)

$$\frac{dz}{dx} = f(x,y,z) \text{ with } y(x_0) = y_0 \text{ and } z(x_0) = z_0$$

The problem reduces to solving the simultaneous equations.

$$\frac{dy}{dx} = z = f_1(x,y,z)$$

$$\frac{dz}{dx} = f_2(x,y,z) \quad \text{subject to } y(x_0) = y_0, \text{ and } z(x_0) = z_0.$$

The predictor and corrector formulae can be written as

$$y_{n+1}^{(P)} = y_{n-3} + \frac{4h}{3}\left[2y'_{n-2} - y'_{n-1} + 2y'_n\right]$$

$$z_{n+1}^{(P)} = z_{n-3} + \frac{4h}{3}\left[2z'_{n-2} - z'_{n-1} + 2z'_n\right]$$

and

$$y_{n+1}^{(c)} = y_{n-1} + \frac{h}{3}\left[y'_{n-1} + 4y'_n + y'_{n+1}\right]$$

$$z_{n+1}^{(c)} = z_{n-1} + \frac{h}{3}\left[z'_{n-1} + 4z'_n + z'_{n+1}\right]$$

In particular when $n = 3$

$$\left.\begin{aligned} y_4^{(P)} &= y_0 + \frac{4h}{3}\left[2y'_1 - y'_2 + 2y'_3\right] \\ z_4^{(P)} &= z_0 + \frac{4h}{3}\left[2z'_1 - z'_2 + 2z'_3\right] \end{aligned}\right\} \qquad (2)$$

and

$$\left.\begin{aligned} y_4^{(c)} &= y_2 + \frac{h}{3}\left[y'_2 + 4y'_3 + y'_4\right] \\ z_4^{(c)} &= z_2 + \frac{h}{3}\left[z'_2 + 4z'_3 + z'_4\right] \end{aligned}\right\} \qquad (3)$$

(2) is called Milne's predictor formula and (3) is called Milne's corrector formula.

Example 1: Given $\frac{d^2y}{dx^2} + x\frac{dy}{dx} + y = 0$, $y(0) = 1$, $y'(0) = 0$ obtain y for x=0, 0.1, 0.2, 0.3 by any method. Further compute $y(0.4)$ by Milne's method.

Solution: The given equation is

$$y'' + xy' + y = 0 \ldots (1) \text{ with } y(0) = 1,\ y'_{(0)} = 1$$

Let $y' = z$ then $y'' = z'$

$$\therefore \quad z' + xz + y = 0 \text{ and } y' = z$$

$$\therefore \quad \frac{dy}{dx} = y' = z = f_1(x,y,z) \text{ with } y(0) = 1$$

$$\frac{dz}{dx} = z' = f_2(x,y,z) = -xz - y \text{ with } z(0) = 0$$

We use Taylor's series method to find y. Differentiating (1) n times we get

$$y_{n+2} + xy_{n+1} + ny_n + y_n = 0$$

At $x = 0$, $(y_{n+2})_0 = -(n+1)y_n$

$$\therefore \quad y(0) = 1 \text{ gives } y_2(0) = -1,\ y_4^{(0)} = 32,\ y_6^{(0)} = -5 \times 3$$

and $y_1(0) = 0$ gives $y_3(0) = y_5(0) = \ldots = 0$.

Expand $y(x)$ by Taylor's series

$$y(x) = y(0) + xy_1(0) + \frac{x^2}{2!}y_2(0) + \frac{x^3}{3!}y_3(0) + \ldots$$

$$\therefore \quad y(x) = 1 - \frac{x^2}{2!} + \frac{3}{4!}x^4 - \frac{5 \times 3}{6!}x^6 + \ldots \tag{2}$$

$$\text{and } z(x) = y'(x) = -x + \frac{1}{2}x^3 - \frac{1}{8}x^5 \ldots = -xy \tag{3}$$

From (2) we have

$$y(0.1) = 1 - \frac{(0.1)^2}{2} + \frac{1}{8}(.1)^4 + \ldots = 0.995$$

$$y(0.2) = 1 - \frac{(0.2)^2}{2} + \frac{(0.2)^4}{8} + \ldots = 0.9802$$

$$y(0.3) = 1 - \frac{(0.3)^2}{2} + \frac{(0.3)^4}{8} - \frac{(0.3)^6}{48} \ldots = 0.956.$$

From (3) we have

$$z(0.1) = -0.0995,\ z(0.2) = -0.196,\ z(0.3) = -0.2863.$$

Also from (1), $z'(x) = -(xz + y)$

$$\therefore \quad z'(0.1) = 0.985,\ z'(0.2) = -0.941,\ z'(0.3) = -0.87.$$

Now, we have to apply Milne's predictor formula first to z and then to y we get

$$z(0.4) = z(0) + \frac{4h}{3}\left[2z'(0.1) - z'(0.2) + 2z'(0.3)\right]$$

$$= 0 + \frac{4}{3}(.1)\left[-1.79 + 0.941 - 1.74\right]$$

$$= -0.3692$$

and

$$y(0.4) = y(0) + \frac{4}{3}(.1)\left[2y'(0.1) - y'(0.2) + 2y'(0.3)\right]$$

$$= 1 + \frac{4}{3}(.1)\left[-0.199 + 0.196 - 0.5736\right]$$

$$= -0.9231.$$

Also $\quad z'(0.4) = -\left[x(0.4)\, z(0.4) + y(0.4)\right]$

$$= \left[0.4(-0.3692) + 0.9232\right]$$

$$= -0.7754.$$

Applying Milne's corrector formula we get

$$z(0.4) = z(0.2) + \frac{h}{3}\left[z'(0.2) + 4z'(0.3) + z'(0.4)\right]$$

$$= -0.196 + \left(\frac{0.1}{3}\right)\left[-0.941 - 3.48 - 0.7754\right]$$

$$= -0.3692$$

and

$$y(0.4) = y(0.2) + \frac{h}{3}\left[y'(0.2) + 4y'(0.3) + y'(0.4)\right]$$

$$= -0.9802 + \left(\frac{0.1}{3}\right)[-0.196 - 1.1452 - 0.3692]$$

$$= 0.9232$$

Hence, $y(0.4) = 0.9232$ and $y(0.4) = -0.3692$.

Example 2: Given $\frac{d^2y}{dx^2} + 2x\frac{dy}{dx} + y = 0$ and $y(0) = 0.5, y'(0) = 0.1$ obtain y for $x = 0, 0.1, 0.2, 0.3$ by any suitable method. Hence, apply Milne's method to compute $y(0.4)$.

Example 3: Given $y'' = xy'^2 - y^2$ and $y(0) = 1, y'(0) = 0$ obtain y for $x = 0,\ 0.1,\ 0.2,\ 0.3$ by Taylor's series method. Hence, apply Milne's method to compute $y(0.4)$.

Ans: $y(0.4) = 0.9232$

2.2.4 Taylor's Series Method

Example 1: Given $y'' + xy' + y = 0, y(0) = 1, y'(0) = 0$. Obtain y for $x = 0,\ 0.1,\ 0.2,\ 0.3$ by Taylor's series method.

Solution: Given second order differential equation is

$$y'' + xy' + y = 0 \tag{1}$$

Let $y' = z$ then $y'' = z'$

$$\therefore \quad z' + xz + y = 0; \quad y' = z \tag{2}$$

We use Taylor's series method to find y.

Differentiating the given equation n times we get

$$y_{n+2} + xy_{n+1} + ny_n + y_n = 0$$

At $x = 0$,

$$(y_{n+2})_0 = -(n+1)y_n$$

$\therefore \quad y(0) = 1$ gives $y_2(0) = -1,\ y_4(0) = 3,\ y_6(0) = -5 \times 3$ and $y_1(0)$ gives $y_3(0) = y_5(0) = \ldots = 0$

Expanding $y(x)$ by Taylor's series, we have

$$y(x) = y(0) + xy_1(0) + \frac{x^2}{2!} y_2(0) + \frac{x^3}{3!} y_3(0) + \ldots$$

and

$$z(x) = y'(x) = 1y_1(0) + xy_2(0) + \frac{x^2}{2} y_3(0) + \ldots$$

$$\therefore \quad y(x) = 1 - \frac{x^2}{2!} + \frac{3}{4!}x^4 - \frac{5 \times 3}{6!}x^6 + \ldots \tag{3}$$

and

$$z(x) = -x + \frac{1}{2}x^3 - \frac{1}{8}x^5 \ldots$$

$$= -x(4) \tag{4}$$

From (2) we have

$$y(0.1) = 1 - \frac{(0.1)^2}{2!} + \frac{1}{8}(0.1)^4 + \ldots = 0.995$$

$$y(0.2) = 1 - \frac{(0.2)^2}{2!} + \frac{1}{8}(0.2)^4 + \ldots = 0.9802$$

$$y(0.3) = 1 - \frac{(0.3)^2}{2!} + \frac{1}{8}(0.3)^4 + \ldots = 0.956$$

From (3)

$$z(0.1) = -0.0995$$

$$z(0.2) = -0.196$$

$$z(0.3) = -0.2863$$

QUESTION BANK

1. Solve the simultaneous first order differential equations $\frac{dy}{dx} = y + 2z$, $\frac{dz}{dx} = 3y + 2z$, given $y = 6$, $z = 4$ at $x = 0$ for y and z at $x = 0.02$ by using fourth order Runge-Kutta method.

2. Using Runge-Kutta method of fourth order find approximate value of y and z corresponding to $x = 0.1$ given that $y(0) = 2,\ z(0) = 1$ and $\frac{dy}{dx} = x + z,\ \frac{dz}{dx} = x - y^2$

   Ans: $y(0.1) = 2.0845,\ z(0.1) = 0.5867.$

3. Solve the differential equations

   $\frac{dy}{dx} = 1 + xz,\ \frac{dz}{dx} = -xy$ for $x = 0.3$ using Picard's method initial values are $x = 0,\ y = 0,\ z = 1.$

   Ans: $y(0.3) = 0.3448,\ z(0.3) = 0.99.$

4. Solve the differential equations

   $\frac{dy}{dx} = yz + x;\ \frac{dz}{dx} = xz + y$ given that $y(0) = 1, z(0) = -1$ for $y(0.2),\ z(0.2)$ using Picard's method.

   Ans: $y(0.2) = 0.85218,\ z(0.2) = -0.8341842$ approximately.

5. Use Runge-Kutta method of order four to solve $\frac{dy}{dx} + xz = 0;$ $\frac{dz}{dx} - y^2 = 0$ at $x = 0.2$ given that $y = 1,\ z = 1$ at $x = 0.$

6. Obtain y for $x = 0.2$, $y'' + 2xy' + y = 0$ and $y = 0.5\ y' = 0.1$ when $x = 0$ by Taylor's series method.

UNIT 3

Functions of Complex Variables – I

3.1 COMPLEX VARIABLE–1

A complex number z is an ordered pair (x, y) of real numbers and is written as

$$z = x + iy \quad \text{where } i = \sqrt{-1}$$

Here x is called the real part of z and y is called imaginary parts of z. In the Arand's diagram, the complex number z is represented by the point $P(x, y)$.

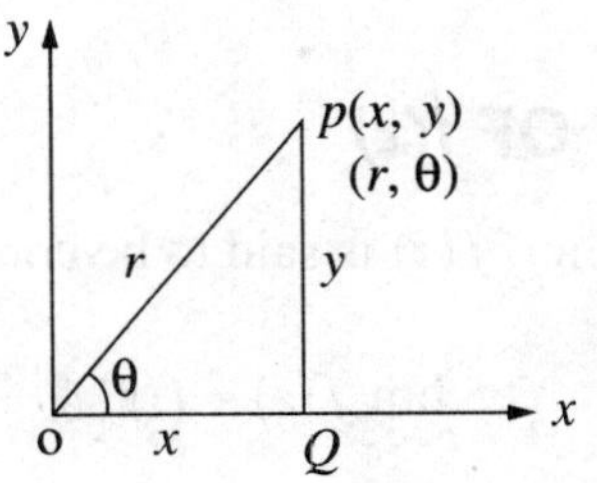

If (r, θ) be the polar coordinates of P then $r = \sqrt{x^2 + y^2}$ and $\theta = \tan^{-1}\dfrac{y}{x}$.

r is called the modulus of z and is denoted by $|z|$ and θ is called the argument of z and is denoted by arg z.

Now,

$$z = x + iy$$

$$= r\cos\theta + ir\sin\theta$$

$$= r\,e^{i\theta} \text{ which is the polar form of complex number}$$

If $z = x + iy$ then $\bar{z} = x - iy$ called the conjugate of the complex number.

$$|\bar{z}| = |z| \text{ and } |z|^2 = z.\bar{z}$$

$$R(z) = \frac{z + \bar{z}}{2} \text{ and } I(z) = \frac{z - \bar{z}}{2}.$$

Definition: If for each value of the complex variable $z(= x + iy)$ in a region R, we have one or more values of $w(= u + iv)$ then w is said to be a function of z and we write

$$w = f(z)$$

$$\text{or} \quad u + iv = f(z)$$

$$\text{or} \quad f(z) = u(x,y) + iv(xy)$$

where u and v are real functions of x and y.

If to each value of z there corresponds one and only value of w, then w is called a single-valued function of z.

If to each value of z there corresponds more than one value of w, then w is called a multi-valued function of z. For example, $w = e^z$ is a single valued function and $w = \pm\sqrt{z}$ is a multi-valued function of z.

3.2 CONTINUITY OF *f(z)*

A single valued function $w = f(z)$ is said to be continuous at $z = z_0$ if

$$\lim_{z \to z_0} f(z) = f(z_0)$$

Further, $f(z)$ is said to be continuous in any region R of the z-plane, if it is continuous at every point of the region.

Also if $w = f(z) = u(x,y) + iv(x,y)$ is continuous at $z = z_0$ then $u(x,y)$ and $V(x,y)$ are also continuous at $z = z_0$, i.e., $x = x_0$, $y = y_0$.

Conversely, if $u(x,y)$ and $v(x,y)$ are continuous at (x_0, y_0) then $f(z)$ will be continuous at $z = z_0$.

3.3 DERIVATIVE OF *f(z)*

Let $w = f(z)$ be a single valued function of $z(= x + iy)$ then the derivative of $w = f(z)$ is defined to be

$$\frac{dw}{dz} = \lim_{\delta z \to 0} \frac{f(z + \delta z) - f(z)}{\delta z} = f'(z).$$

Provided the limit exists and has the same value for all the different ways in which $\delta z \to 0$

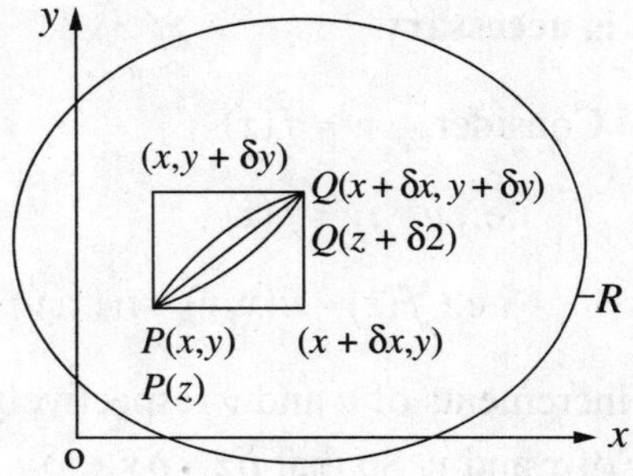

Suppose $P(z)$ is fixed an $\theta(z + \delta z)$ is a neighbouring point. The point Q may approach P along any straight line or curved path in the given region, i.e., δz may tend to zero in any manner.

3.4 ANALYTIC FUNCTION

A function $f(z)$ which is single valued and possesses a unique derivative with respect to z at all points of a region R is called an analytic or a regular function of z in that region.

A point at which an analytic function ceases to possess a derivative is called a singular point of the function.

3.5 NECESSARY AND SUFFICIENT CONDITIONS FOR *w = f(z)* TO BE ANALYTIC

The necessary and sufficient conditions for the function

$$w = f(z) = u(xy) + iv(xy)$$

to be analytic in a region R, are

i) $\dfrac{\partial u}{\partial x}, \dfrac{\partial u}{\partial y}, \dfrac{\partial v}{\partial x}, \dfrac{\partial v}{\partial y}$ are continuous functions of x and y in the region R

ii) $\dfrac{\partial u}{\partial x} = \dfrac{\partial v}{\partial y}; \quad \dfrac{\partial u}{\partial y} = -\dfrac{\partial v}{\partial x}.$

The conditions (ii) are known as Cauchy-Riemann equations or briefly $C - R$ equations.

Proof. **(a) Condition is necessary**

$$\text{Consider,} \quad w = f(z)$$

$$\text{i.e., } u + iv = f(z)$$

$$\text{i.e., } f(z) = u(x,y) + iv(x,y)$$

Let δu and δv be the increments of u and v respectively, corresponding to the increments δx and δy of x and y. So that $\delta z = \delta x + i\delta y$ $(\because \quad z = x + iy)$

If $f(z)$ possesses a unique derivative at $P(z)$ then

$$f'(z) = \lim_{\delta z \to 0} \frac{f(z+\delta z) - f(z)}{\delta z}$$

$$= \lim_{\delta z \to 0} \frac{[(u+\delta u) + i(v+\delta v) - (u+iv)]}{\delta z}$$

$$= \lim_{\delta z \to 0} \left(\frac{\delta u}{\delta z} + i\frac{\delta v}{\delta z}\right) \ldots \tag{A}$$

Since δz can approach zero in any manner, we can first assume that δz is wholly real and then wholly imaginary.

when δz is wholly real, i.e., $\delta z = \delta x$ and $\delta y = 0$

$$\text{From (A), } f'(z) = \lim_{\delta x \to 0} \left(\frac{\delta u}{\delta x} + i\frac{\delta v}{\delta x}\right)$$

$$= \frac{\partial u}{\partial x} + i\frac{\partial v}{\partial x} \tag{1}$$

when δz is wholly imaginary $\delta x = 0$, and $\delta z = i\delta y$

From (A)

$$f'(z) = \lim_{i\delta y \to 0} \left(\frac{\delta u}{i\delta y} + \frac{i\delta r}{i\delta y}\right)$$

$$= \frac{1}{i}\frac{\partial u}{\partial y} + \frac{\partial v}{\partial y}$$

$$= -i\frac{\partial u}{\partial y} + \frac{\partial v}{\partial y} \ldots \quad (\because \quad i^2 = -1) \tag{2}$$

Now the existence of $f'(z)$ requires the equality of (i) and (ii)

$$\therefore \quad \frac{\partial u}{\partial x} + i\frac{\partial v}{\partial x} = \frac{\partial v}{\partial y} - i\frac{\partial u}{\partial y}$$

on equating real and imaginary parts we get

$$\boxed{\frac{\partial u}{\partial x}=\frac{\partial v}{\partial y};\qquad \frac{\partial u}{\partial y}=-\frac{\partial v}{\partial x}}$$

Thus for the necessary conditions for the existence of the derivative of $f(z)$ is that $C-R$ equations should be satisfied.

3.5.1 Condition is Sufficient

Suppose $f(z)$ is a single valued-function possessing partial derivatives $\frac{\partial u}{\partial x}$, $\frac{\partial u}{\partial y}$, $\frac{\partial v}{\partial x}$, $\frac{\partial v}{\partial y}$ at each point of the region R and that the $C-R$ equations are satisfied.

$$\text{i.e., } \frac{\partial u}{\partial x}=\frac{\partial v}{\partial y};\ \frac{\partial u}{\partial y}=-\frac{\partial v}{\partial x}$$

We shall show that $f(z)$ is analytic, i.e., $f'(z)$ exists at every point of the region R.

By Taylor's theorem for functions of two variables. We have

$$f(z+\delta z)=u(x+\delta x,\ y+\delta y)+iv(x+\delta x,\ y+\delta y)$$

$$=\left[u(x,y)+\left(\frac{\partial u}{\partial x}\,\delta x+\frac{\partial u}{\partial y}\,\delta y\right)\right]+i\left[v(x,y)+\left(\frac{\partial v}{\partial x}\delta x+\frac{\partial v}{\partial y}\right)\delta y\right]+\ldots$$

Omitting the terms beyond the first powers of δx and δy.

$$=[u(x,y)+iv(x,y)]+\left(\frac{\partial u}{\partial x}+i\frac{\partial v}{\partial x}\right)\delta x+\left(\frac{\partial u}{\partial y}+i\frac{\partial v}{\partial y}\right)\delta y$$

$$=f(z)+\left(\frac{\partial u}{\partial x}+i\frac{\partial v}{\partial x}\right)\delta x+\left(\frac{\partial u}{\partial y}+i\frac{\partial v}{\partial y}\right)\delta y$$

$$\therefore\quad f(z+\delta z)-f(z)=\left(\frac{\partial u}{\partial x}+i\frac{\partial v}{\partial x}\right)\delta x+\left(-\frac{\partial v}{\partial x}+i\frac{\partial u}{\partial x}\right)\delta y$$

($\because$ by $-CR$ equations)

$$=\left(\frac{\partial u}{\partial x}+i\frac{\partial v}{\partial x}\right)\delta x+\left(i^2\frac{\partial v}{\partial x}+i\frac{\partial u}{\partial x}\right)\delta y\qquad \because i^2=-1$$

$$=\left(\frac{\partial u}{\partial x}+i\frac{\partial v}{\partial x}\right)(\delta x+i\delta y)$$

$$=\left(\frac{\partial u}{\partial x}+i\frac{\partial v}{\partial x}\right)\delta z\qquad (\because\ \delta z=\delta x+i\delta y)$$

$$\therefore \quad \frac{f(z+\delta z)-f(z)}{\delta z}=\frac{\partial u}{\partial x}+i\frac{\partial v}{\partial x}$$

$$\therefore \quad f'(z)=\lim_{\delta z\to 0}\frac{f(z+\delta z)-f(z)}{\delta z}=\frac{\partial u}{\partial x}+i\frac{\partial v}{\partial x}.$$

Thus $f'(z)$ exists and hence $f(z)$ is analytic.

3.6 CONJUGATE FUNCTIONS

The real and imaginary parts of an analytic function are called conjugate functions. The relation between two conjugate functions is given by $C-R$ equations.

3.7 CAUCHY RIEMANN EQUATIONS IN POLAR COORDINATES

If $f(z)=u(r,\theta)+iv(r,\theta)$ is analytic then prove that

$$\frac{\partial u}{\partial r}=\frac{1}{r}\frac{\partial v}{\partial \theta};\quad \frac{\partial v}{\partial r}=-\frac{1}{r}\frac{\partial u}{\partial \theta}.$$

Proof. Let (r,θ) be the polar coordinates of the point whose cartesian co-ordinates are (x,y) then

$$x=r\cos\theta,\ y=r\sin\theta$$

$$\because \quad z=x+iy=r(\cos\theta+i\sin\theta)=r\,e^{i\theta}$$

Now,

$$u+iv=f(z)=f(r\,e^{i\theta}) \tag{1}$$

Differentiating (1) partially w.r.t. r, we get

$$\frac{\partial u}{\partial r}+i\frac{\partial v}{\partial r}=f'(r\,e^{i\theta})e^{i\theta} \tag{2}$$

Differentiating (1) partially w.r.t. θ we get

$$\frac{\partial u}{\partial \theta}+i\frac{\partial v}{\partial \theta}=f'(r\,e^{i\theta})\,r\,ie^{i\theta}$$

$$=ir\left(\frac{\partial u}{\partial r}+i\frac{\partial v}{\partial r}\right) \quad \text{from (2)}$$

$$= -r\frac{\partial v}{\partial r} + ir\frac{\partial u}{\partial r}$$

Equating real and imaginary parts we get

$$\frac{\partial u}{\partial \theta} = -r\frac{\partial v}{\partial r}; \quad \frac{\partial v}{\partial \theta} = r\frac{\partial u}{\partial r}.$$

$$\therefore \quad \boxed{\frac{\partial u}{\partial r} = \frac{1}{r}\frac{\partial v}{\partial \theta}; \quad \frac{\partial v}{\partial r} = -\frac{1}{r}\frac{\partial u}{\partial \theta}}$$ which are $C - R$ equations in polar form.

3.8 HARMONIC FUNCTIONS

Definition: Any solution of the Laplace equation, $\frac{\partial^2 \psi}{\partial x^2} + \frac{\partial^2 \psi}{\partial y^2} = 0$ is called a harmonic function.

Theorem: If $f(z) = u + iv$ is an analytic function then show that u and v are both harmonic.

Proof. If $f(z) = u + iv$ is an analytic function in some region of the z-plane then $C - R$ equations are satisfied.

$$\therefore \quad \frac{\partial u}{\partial x} = \frac{\partial v}{\partial y} \tag{1}$$

$$\text{and} \quad \frac{\partial u}{\partial y} = -\frac{\partial v}{\partial x} \tag{2}$$

Differentiating (1) partial w.r.t. x and (2) w.r.t. y we get

$$\frac{\partial^2 u}{\partial x^2} = \frac{\partial^2 v}{\partial x\,\partial y} \tag{3}$$

$$\frac{\partial^2 u}{\partial y^2} = -\frac{\partial^2 v}{\partial y\,\partial x} \tag{4}$$

Since $\frac{\partial^2 v}{\partial x\,\partial y} = \frac{\partial^2 v}{\partial y\,\partial x}$ and adding (3) and (4) we get

$$\frac{\partial^2 u}{\partial x^2} + \frac{\partial^2 u}{\partial y^2} = 0 \tag{5}$$

Now, differentiating (1) w.r.t. y and (2) w.r.t. x partially and subtracting we get

$$\frac{\partial^2 v}{\partial x_2} + \frac{\partial^2 v}{\partial y_2} = 0 \tag{6}$$

Equations (5) and (6) show that the real and imaginary parts u and v of an analytic function satisfying the Laplace's equation.

Hence, u and v are known as harmonic functions.

3.9 ORTHOGONAL SYSTEM

Every analytic function $f(z) = u + iv$ defines two families of curves $u(x,y) = c_1$ and $v(x,y) = c_2$ which form an orthogonal system.

Consider the two families of curves

$$u(x,y) = c_1 \quad (1) \quad \text{and} \quad v(x,y) = c_2 \quad (2)$$

Differentiating (1) w.r.t. x, we get

$$\frac{\partial u}{\partial x} + \frac{\partial u}{\partial y}\frac{dy}{dx} = 0 \quad \text{or} \quad \frac{dy}{dx} = -\frac{\dfrac{\partial u}{\partial x}}{\dfrac{\partial u}{\partial y}} = m_1 \text{ (say).}$$

Similarly from (2)

$$\frac{dy}{dx} = \frac{-\dfrac{\partial v}{\partial x}}{\dfrac{\partial v}{\partial y}} = m_2 \text{ (Say)}$$

$$\therefore\ m_1 m_2 = \frac{\dfrac{\partial u}{\partial x}\dfrac{\partial v}{\partial x}}{\dfrac{\partial u}{\partial y}\dfrac{\partial v}{\partial y}} = \frac{\dfrac{\partial v}{\partial y}\dfrac{\partial v}{\partial x}}{-\dfrac{\partial v}{\partial y}\dfrac{\partial v}{\partial y}} = -1 \left(\because \text{ by } C\text{–}R \text{ eqns. } \frac{\partial u}{\partial x} = \frac{\partial v}{\partial y}, \frac{\partial u}{\partial y} = -\frac{\partial v}{\partial x}\right)$$

Thus, the product of the slopes of the curves (1) and (2) is -1. Hence, the curves intersect at right angles, i.e., they form an orthogonal system.

3.9.1 Application of Analytic Functions to Flow Problems

Let $f(z) = u + iv$ be an analytic function. The real and imaginary parts of an analytic function satisfy Laplace equation in two variables. These conjugate functions provide solutions to a number of field and flow problems.

Example:

Consider the two-dimensional irrotational motion of an incompressible fluid, in planes parallel to xy-plane.

Let V be the velocity of a fluid particle, then it can be expressed as

$$V = v_x i + v_y j \tag{1}$$

Since the motion is irrotational, there exists a scalar function $\phi(x,y)$ such that

$$V = \nabla\phi = \frac{\partial\phi}{\partial x} i + \frac{\partial\phi}{\partial y} j \tag{2}$$

From (1) and (2) we have

$$V_x = \frac{\partial\phi}{\partial x} \quad \text{and} \quad V_y = \frac{\partial\phi}{\partial y} \tag{3}$$

The scalar function $\phi(x,y)$, which gives the velocity components, is called the velocity potential function or simply the *velocity potential.*

Also the fluid is incompressible, div $V = 0$ i.e. $\nabla \cdot \overline{V} = 0$

$$\text{i.e.,} \left(i\frac{\partial}{\partial x} + j\frac{\partial}{\partial y} \right) \cdot (v_x i + v_y j) = 0$$

$$\frac{\partial v_x}{\partial x} + \frac{\partial v_y}{\partial y} = 0 \tag{4}$$

Substituting the values of V_x and V_y in (4)

$$\frac{\partial}{\partial x}\left(\frac{\partial\phi}{\partial x}\right) + \frac{\partial}{\partial y}\left(\frac{\partial\phi}{\partial y}\right) = 0$$

$$\text{or} \quad \frac{\partial^2\phi}{\partial x^2} + \frac{\partial^2\phi}{\partial y^2} = 0$$

Thus, the function ϕ is harmonic and can be treated as real part of an analytic function

$$w = f(z) = \phi(x,y) + i\psi(x,y)$$

For interpretation of conjugate function $\psi(x,y)$ the slope at any point of the curve $\psi(x,y) = C_1$ is given by

$$\frac{dy}{dx} = -\frac{-\dfrac{\partial\psi}{\partial x}}{\dfrac{\partial\psi}{\partial y}} = \frac{\dfrac{\partial\phi}{\partial y}}{\dfrac{\partial\phi}{\partial x}} \quad \text{by } C-R \text{ equations}$$

$$= \frac{V_y}{V_x}.$$

This shows that the resultant velocity $\sqrt{v_x^2 + v_y^2}$ of the fluid particle is along the tangent to the curve $\psi(x,y) = \gamma$, i.e., the fluid particles move along this curve such curves are known as *stream lines* and $\psi(x,y)$ is called the *stream function*. The curves represented by $\phi(x,y) = c$ are called equipotential lines.

Since $\phi(x,y)$ and $\psi(x,y)$ are conjugate functions of analytic function $w = f(z)$, the equipotential lines $\phi(x,y) = c$ and the stream lines $\psi(xy) = \gamma$ intersect each other orthogonally.

$$\text{Now,} \quad \frac{dw}{dz} = \frac{\partial \phi}{\partial x} + i\frac{\partial \psi}{\partial x}$$

$$= \frac{\partial \phi}{\partial x} - i\frac{\partial \phi}{\partial y} \quad \text{(by } C-R \text{ eqns.)}$$

$$= v_x - iv_y$$

$\therefore$ The magnitude of the resultant velocity $= \left|\frac{dw}{dz}\right| = \sqrt{{v_x}^2 + {v_y}^2}$.

The function $w = f(z)$ which fully represents the flow pattern is called the *complex potential.*

In heat flow problems, the curves $\phi(x,y) = c$ and $\psi(x,y) = \gamma$ are known as isothermals and heat flow lines respectively.

Before we proceed to the problems recollect the following some of the formulae.

1. $\sin(ix) = i\sinh x$
2. $\cos(ix) = \cosh x$
3. $\sin z = \sin(x+iy)$
   $= \sin x \cos(iy) + \cos x \sin(iy)$
   $= \sin x \cosh y + i\cos x \sinh y$
4. $\cos z = \cos(x+iy)$
   $= \cos x \cos(iy) - \sin x \sin(iy)$
   $= \cos x \cosh y - i\sin x \sinh y$
5. $e^z = e^{x+iy} = e^x \cdot e^{iy}$
   $= e^x(\cos y + i\sin y)$
   $= e^x \cos y + ie^x \sin y$

6. $\log z = \log(x + iy)$

$= \log(r\cos\theta + ir\sin\theta)$

$= \log^{r\,e^{i\theta}} = \log^{r} + i\theta$

$= \frac{1}{2}\log(x^2 + y^2) + i\tan^{-1}\frac{y}{x}$

7. $|z| = \sqrt{x^2 + y^2}$

8. $\frac{e^z - e^{-z}}{2} = \sinh z, \quad \frac{e^z + e^{-z}}{2} = \cosh z$

9. $(\cos\theta + i\sin\theta)^n = \cos n\theta + i\sin n\theta$

10. $(\cos\theta + i\sin n\theta)^{-n} = \cos n\theta - i\sin n\theta$

Example 1: Show that the function $f(z) = xy + iy$ is everywhere continuous but is not analytic.

Solution: Let $f(z) = u + iv = xy + iy$ be the given function.

$$\therefore \quad u = xy, \qquad v = y$$

$$\frac{\partial u}{\partial x} = y; \qquad \frac{\partial v}{\partial x} = 0$$

$$\frac{\partial u}{\partial y} = x; \qquad \frac{\partial v}{\partial y} = 1$$

Thus, $\frac{\partial u}{\partial x} \neq \frac{\partial v}{\partial y}; \ \frac{\partial u}{\partial y} \neq -\frac{\partial v}{\partial x}.$

$\therefore$ $C-R$ equations are not satisfied anywhere and we see that both u and v being polynomials in x and y are continuous everywhere.

$\therefore$ $f(z)$ is continuous everywhere.

Hence, $f(z)$ is not analytic at any point but is continuous everywhere.

Example 2: Show that $w = \log^z$ is analytic and find $\frac{dw}{dz}$.

Solution: Given $w = \log^z$

$$\therefore \quad u + iv = \log(x + iy)$$

$$= \frac{1}{2}\log(x^2 + y^2) + i\tan^{-1}\frac{y}{x}$$

$$\therefore \quad u = \frac{1}{2}\log(x^2 + y^2); \qquad v = \tan^{-1}\frac{y}{x}$$

$$\frac{\partial u}{\partial x} = \frac{x}{x^2 + y^2}; \qquad \frac{\partial v}{\partial x} = \frac{-y}{x^2 + y^2}$$

$$\frac{\partial u}{\partial y} = \frac{y}{x^2 + y^2}; \qquad \frac{\partial v}{\partial y} = \frac{x}{x^2 + y^2}$$

$\therefore$ $C - R$ equations are satisfied and the partial derivatives are continuous, except at $(0, 0)$. Hence, $w = \log z$ is analytic.

Now,

$$f'(z) = \frac{dw}{dz} = \frac{\partial u}{\partial x} + i\frac{\partial v}{\partial x}$$

$$= \frac{x}{x^2 + y^2} + i\left(\frac{-y}{x^2 + y^2}\right)$$

$$= \frac{x - iy}{(x + iy)(x - iy)}$$

$$= \frac{1}{x + iy} = \frac{1}{z}$$

Thus, if $w = \log^z$ then $\dfrac{dw}{dz} = \dfrac{1}{z}$.

Example 3: Show that $w = e^z$ is an analytic function and determine $f'(z)$.

Solution: Given $w = e^z$

$$\therefore \quad u + iv = e^{x + iy} = e^x \cdot e^{iy}$$

$$= e^x \cos y + ie^x \sin y$$

$$\therefore \quad u = e^x \cos y, \qquad v = e^x \sin y$$

$$\therefore \quad \frac{\partial u}{\partial x} = e^x \cos y, \quad \frac{\partial v}{\partial x} = e^x \sin y$$

$$\frac{\partial u}{\partial y} = -e^x \sin y, \quad \frac{\partial v}{\partial y} = e^x \cos y$$

$\therefore$ $C - R$ equations are satisfied and $w = e^z$ is an analytic function.

Now,

$$f'(z) = \frac{\partial u}{\partial x} + i\frac{\partial v}{\partial x}$$

$$= e^x \cos y + ie^x \sin y$$

$$= e^x(\cos y + i \sin y)$$

$$= e^x \cdot e^{iy} = e^{x+iy} = e^z$$

Thus, if $w = e^z$ then $\dfrac{dw}{dz} = f'(z) = e^z$.

Example 4: If $w = \sin z$ then show that w is an analytic function and find $\dfrac{dw}{dz}$.

Solution: Given $w = \sin z$

$$\therefore \quad u + iv = \sin(x + iy)$$

$$= \sin x \cosh y + i \cos x \sinh y$$

$$\therefore \quad u = \sin x \cosh y \qquad v = \cos x \sinh y$$

$$\therefore \quad \frac{\partial u}{\partial x} = \cos x \cosh y \qquad \frac{\partial v}{\partial x} = -\sin x \sinh y$$

$$\frac{\partial u}{\partial y} = \sin x \sinh y \qquad \frac{\partial v}{\partial y} = \cos x \cosh y$$

$\therefore$ $C - R$ equations are satisfied and hence $w = \sin z$ is an analytic function.

Now,

$$f'(z) = \frac{dw}{dz} = \frac{\partial u}{\partial x} + i\frac{\partial v}{\partial x}$$

$$= \cos x \cosh y - i \sin x \sinh y$$

$$= \cos x \cos(iy) - \sin x \sin(iy)$$

$$= \cos(x + iy)$$

$$= \cos z.$$

Thus, if $w = \sin z$ then $\dfrac{dw}{dz} = \cos z$.

Example 5: If $w = \cos z$ find $\dfrac{dw}{dz}$ show that w is an analytic function.

Solution: Given $w = \cos z$.

$$\therefore \quad u + iv = \cos x \cosh y - i \sin x \sinh y$$

$$\therefore \quad u = \cos x \cosh y, \qquad v = -\sin x \sinh y$$

$$\frac{\partial u}{\partial x} = -\sin x \cosh y, \qquad \frac{\partial v}{\partial x} = -\cos x \sinh y$$

$$\frac{\partial u}{\partial y} = \cos x \sinh y, \qquad \frac{\partial v}{\partial y} = -\sin x \cosh y$$

$\therefore C-R$ equations are satisfied and hence $w = \cos z$ is analytic.

$$\text{Now,} \quad f'(2) = \frac{dw}{dz} = \frac{\partial u}{\partial x} + i\frac{\partial v}{\partial x}$$

$$= -\sin x \cosh hy + i \cos x \sin hy$$

$$= -[\sin x \cos(iy) + \cos x \sin(iy)]$$

$$= -\sin(x + iy)$$

$$= -\sin z.$$

Thus, if $w = \cos z$ then $\dfrac{dw}{dz} = -\sin z.$

Example 6: Show that the function $f(z) = z^2$ is analytic for all z and $f'(z) = 2z.$

Solution: Given $f(z) = z^2$

$$= (x + iy)^2 = x^2 - y^2 + 2ixy$$

$$\therefore \quad u + iv = (x^2 - y^2) + i(2xy)$$

$$\therefore \quad u = x^2 - y^2, \qquad v = 2xy$$

$$\frac{\partial u}{\partial x} = 2x \qquad \frac{\partial v}{\partial x} = 2y$$

$$\frac{\partial u}{\partial y} = -2y \qquad \frac{\partial v}{\partial y} = 2x.$$

$\therefore$ $C-R$ equations are satisfied for all x and y. Hence, the function $f(z) = z^2$ is analytic.

$$\text{Now,} \quad f'(z) = \frac{\partial u}{\partial x} + i\frac{\partial v}{\partial x}$$

$$= 2x + i2y = 2(x + iy)$$

$$= 2z$$

Thus if $f(z) = z^2$ then $f'(z) = 2z$.

Example 7: If $w = z^n$ is an analytic function then show that $\dfrac{dw}{dz} = nz^{n-1}$

Solution: Given $w = z^n$

$$\therefore \quad u + iv = (x + iy)^n$$

$$= (r\cos\theta + ir\sin\theta)^n$$

$$= r^n(\cos n\theta + i\sin n\theta)$$

$$\therefore \quad u = r^n \cos n\theta, \qquad v = r^n \sin n\theta.$$

$$\frac{\partial u}{\partial r} = nr^{n-1}\cos n\theta, \qquad \frac{\partial v}{\partial r} = n\,r^{n-1}\sin n\theta$$

$$\frac{\partial u}{\partial \theta} = -\,nr^n \sin n\theta, \qquad \frac{\partial v}{\partial \theta} = nr^n \cos n\theta$$

$$\therefore \quad \frac{\partial u}{\partial r} = \frac{1}{r}\frac{\partial r}{\partial \theta}; \qquad \frac{\partial v}{\partial r} = -\frac{1}{r}\frac{\partial u}{\partial \theta}.$$

$\therefore$ $C - R$ equations in polar form are satisfied and therefore $w = z^n$ is an analytic function.

Now, $$\frac{\partial u}{\partial r} + i\frac{\partial v}{\partial r} = f'(re^{i\theta}).e^{i\theta}.$$

$$\therefore \quad f'(re^{i\theta}) = \frac{1}{ei\theta}\left(\frac{\partial u}{\partial r} + i\frac{\partial v}{\partial r}\right)$$

$$= e^{-i\theta}\left[nr^{n-1}\cos n\theta \sin r^{n-1}\sin n\theta\right]$$

$$= n\,e^{-i\theta} r^{n-1}(\cos n\theta + i\sin n\theta)$$

$$= n\,r^{n-1} e^{-i\theta} e^{in\theta}$$

$$= n\,r^{n-1} e^{i(n-1)\theta}$$

$$= n\left(r^{n-1}\right)\left(e^{i\theta}\right)^{n-1}$$

$$= n(r^{n-1})(\cos\theta + i\sin\theta)^{n-1}$$

$$= n(r\cos\theta + i\sin\theta)^{n-1}$$

$$= n(x+iy)^{n-1}$$

$$= nz^{n-1}$$

Thus, if $w = z^n$ then $\dfrac{dw}{dz} = nz^{n-1}$.

Example 8: If $f(z) = u(r,\theta) + iv(r,\theta)$ is an analytic function show that u and v satisfy the equation

$$\frac{\partial^2 u}{\partial r^2} + \frac{1}{r}\frac{\partial u}{\partial r} + \frac{1}{r^2}\frac{\partial^2 u}{\partial \theta^2} = 0$$

Solution: We have by $C-R$ equations in polar form

$$\frac{\partial v}{\partial \theta} = r\frac{\partial u}{\partial r}, \quad \frac{\partial v}{\partial r} = -\frac{1}{r}\frac{\partial u}{\partial \theta} \qquad (1)$$

Differentiating the first w.r.t 'r' and second w.r.t θ.

$$\therefore \quad \frac{\partial^2 v}{\partial r \partial \theta} = r\frac{\partial^2 u}{\partial r^2} + \frac{\partial u}{\partial r}, \quad \frac{\partial^2 v}{\partial \theta \partial r} = -\frac{1}{r}\frac{\partial^2 u}{\partial \theta^2}$$

Since $\dfrac{\partial^2 r}{\partial r \partial \theta} = \dfrac{\partial^2 r}{\partial \theta \partial r}$ we have

$$r\frac{\partial^2 u}{\partial r^2} + \frac{\partial u}{\partial r} = -\frac{1}{r}\frac{\partial^2 u}{\partial \theta^2}$$

$$\therefore \quad \frac{\partial^2 u}{\partial r^2} + \frac{1}{r}\frac{\partial u}{\partial r} + \frac{1}{r^2}\frac{\partial^2 u}{\partial \theta^2} = 0.$$

Example 9: Show that an analytic function with constant modulus is constant.

Solution: Let an analytic function be $f(z) = u + iv$.

Given $|f(z)| = c$ (constant, say)

$$\text{i.e.} \quad u^2 + v^2 = c^2$$

$$\therefore \quad 2u\frac{\partial u}{\partial x} + 2v\frac{\partial v}{\partial x} = 0$$

$$\therefore \quad u\frac{\partial u}{\partial x} + v\frac{\partial v}{\partial x} = 0 \qquad (1)$$

and similarly,

$$u\frac{\partial u}{\partial y} + v\frac{\partial v}{\partial y} = 0$$

$$-u\frac{\partial v}{\partial x} + v\frac{\partial u}{\partial x} = 0 \quad \text{(by } C\text{-}R \text{ equations)} \qquad (2)$$

Squaring and adding (1) and (2)

$$(u^2 + v^2)\left\{\left(\frac{\partial u}{\partial x}\right)^2 + \left(\frac{\partial v}{\partial x}\right)^2\right\} = 0$$

$$\therefore \quad \left(\frac{\partial u}{\partial x}\right)^2 + \left(\frac{\partial v}{\partial x}\right)^2 = 0 \qquad \because \quad u^2 + v^2 = c^2 \neq 0$$

$$\therefore \quad |f'(z)|^2 = 0$$

$$\therefore \quad f'(z) = 0$$

Hence, $f(z)$ is a constant function.

Example 10: If $f(z)$ is a regular function of z prove that

$$\left(\frac{\partial^2}{\partial x^2} + \frac{\partial^2}{\partial y^2}\right)|f(z)|^2 = 4|f'(z)|^2$$

Solution: Let $f(z) = u(x,y) + iv(x,y)$ so that

$$|f(z)|^2 = u^2 + v^2 = \phi(x,y), \text{ (say)}$$

i.e., $\phi = u^2 + v^2$

$$\therefore \quad \frac{\partial \phi}{\partial x} = 2u\frac{\partial u}{\partial x} + 2v\frac{\partial v}{\partial x}$$

and $$\frac{\partial^2 \phi}{\partial x^2} = 2\left[u\frac{\partial^2 u}{\partial x^2} + \left(\frac{\partial u}{\partial x}\right)^2 + v\frac{\partial^2 v}{\partial x^2} + \left(\frac{\partial v}{\partial x}\right)^2\right] \qquad (1)$$

Similarly,

$$\frac{\partial^2 \phi}{\partial y^2} = 2\left[u\frac{\partial^2 u}{\partial y^2} + \left(\frac{\partial u}{\partial y}\right)^2 + v\frac{\partial^2 v}{\partial y^2} + \left(\frac{\partial v}{\partial y}\right)^2\right] \qquad (2)$$

Adding (1) and (2) we get

$$= 2\left[u\left(\frac{\partial^2 u}{\partial x^2} + \frac{\partial^2 u}{\partial y^2}\right) + v\left(\frac{\partial^2 v}{\partial x^2} + \frac{\partial^2 v}{\partial y^2}\right)\right]$$

$$+ 2\left[\left(\frac{\partial u}{\partial x}\right)^2 + \left(\frac{\partial v}{\partial x}\right)^2 + \left(\frac{\partial u}{\partial y}\right)^2 + \left(\frac{\partial v}{\partial y}\right)^2\right]$$

Since u and v satisfy Laplace equation and C–R equations.

i.e., $\frac{\partial^2 u}{\partial x^2} + \frac{\partial^2 u}{\partial y^2} = 0$, $\frac{\partial^2 u}{\partial x^2} + \frac{\partial^2 v}{\partial y^2} = 0$ and $\frac{\partial u}{\partial x} = +\frac{\partial v}{\partial y}; \frac{\partial u}{\partial y} = -\frac{\partial v}{\partial x}$ we get

$$= 2\left[\left(\frac{\partial u}{\partial x}\right)^2 + \left(\frac{\partial v}{\partial x}\right)^2 + \left(\frac{-\partial v}{\partial x}\right)^2 + \left(\frac{\partial u}{\partial x}\right)^2\right]$$

$$= 4\left[\left(\frac{\partial u}{\partial x}\right)^2 + \left(\frac{\partial v}{\partial x}\right)^2\right]$$

$$\therefore \quad \left(\frac{\partial^2}{\partial x^2} + \frac{\partial^2}{\partial y^2}\right)\phi = 4|f'(z)|^2$$

$$\therefore \quad \left(\frac{\partial^2}{\partial x^2} + \frac{\partial^2}{\partial y^2}\right)|f(z)|^2 = 4|f'(z)|^2.$$

Example 11: If $f(z)$ is a analytic function of z prove that

$$\left(\frac{\partial^2}{\partial x^2} + \frac{\partial^2}{\partial y^2}\right)|Rf(z)|^2 = 2|f'(z)|^2$$

Solution: Let $f(z) = u + iv$ so that $Rf(z) = u$

$$\therefore \quad |Rf(z)|^2 = u^2$$

$$\therefore \quad \frac{\partial^2}{\partial x^2}(u^2) = 2\left[u\frac{\partial^2 u}{\partial x^2} + \left(\frac{\partial u}{\partial x}\right)^2\right] \quad (1)$$

$$\text{and} \quad \frac{\partial^2}{\partial y^2}(u^2) = 2\left[u\frac{\partial^2 u}{\partial y^2} + \left(\frac{\partial u}{\partial y}\right)^2\right] \quad (2)$$

Adding (1) and (2) we get

$$\left(\frac{\partial^2}{\partial x^2} + \frac{\partial^2}{\partial y^2}\right)(u^2) = 2\left[u\left(\frac{\partial^2 u}{\partial x^2} + \frac{\partial^2 u}{\partial y^2}\right) + \left(\frac{\partial u}{\partial x}\right)^2 + \left(\frac{\partial u}{\partial y}\right)^2\right]$$

$$= 2\left[\left(\frac{\partial u}{\partial x}\right)^2 + \left(\frac{\partial v}{\partial x}\right)^2\right] \quad \because \quad \frac{\partial^2 u}{\partial x^2} + \frac{\partial^2 v}{\partial y^2} = 0 \text{ and by } C\text{-}R \text{ eqns.}$$

$$= 2|f'(z)|^2$$

$$\therefore \quad \left(\frac{\partial^2}{\partial x^2} + \frac{\partial^2}{\partial y^2}\right)|Rf(z)|^2 = 2|f'(z)|^2.$$

3.9.2 Milne Thomson Method

1. When u is given

$$f'(z) = \phi_1(z,0) - i\phi_2(z,0)$$

where $\phi_1(z,0) = \left(\frac{\partial u}{\partial x}\right)_{\substack{x=z\\y=0}}$ and $\phi_2(z,0) = \left(\frac{\partial u}{\partial y}\right)_{\substack{x=z\\y=0}}$.

2. When v is given

$$f'(z) = \psi_1(z,0) + i\psi_2(z,0)$$

where $\psi(z,0) = \left(\frac{\partial v}{\partial x}\right)_{\substack{x=z\\y=0}}$ and $\psi_2(z,0) = \left(\frac{\partial v}{\partial y}\right)_{\substack{x=z\\y=0}}$.

Example 12: Prove that $u = y^3 - 3x^2y$ is a harmonic function. Determine its harmonic conjugate and find the corresponding analytic function $f(z)$ interms of z.

Solution: Given $u = y^3 - 3x^2y$

$$\therefore \quad \frac{\partial u}{\partial x} = -6xy \qquad \frac{\partial u}{\partial y} = 3y^2 - 3x^2$$

$$\frac{\partial^2 u}{\partial x^2} = -6y \qquad \frac{\partial^2 u}{\partial y^2} = 6y$$

$$\therefore \quad \frac{\partial^2 u}{\partial x^2} + \frac{\partial^2 u}{\partial y^2} = -6y + 6y$$

$$\therefore \quad \frac{\partial^2 u}{\partial x^2} + \frac{\partial^2 u}{\partial y^2} = 0$$

Since u satisfies Laplace equation, it is a harmonic function.

Let v be the harmonic conjugate to u.

$$\therefore \quad dv = \frac{\partial v}{\partial x}dx + \frac{\partial v}{\partial y}dy$$

$$= -\frac{\partial u}{\partial y}dx + \frac{\partial u}{\partial x}dy \qquad \text{(by } C-R \text{ equations)}$$

$$= -(3y^2 - 3x^2)dx + (-6xy)dy$$

Integrating

$$v = -3xy^2 + x^3 - 3xy^2 + c$$

$$= x^3 - 6xy^2 + c \text{ is the harmonic conjugate to } u.$$

Now, the analytic function $f(z)$ is given by

$$f(z) = u + iv$$

$$= y^3 - 3x^2y + i(-3xy^2 + x^3 + c)$$

$$= i(x + iy)^3 + ic$$

$$= iz^3 + c \qquad \text{where } c \text{ is complex constant.}$$

OR

By Milne's method

$$\phi_1(z,0) = \left(\frac{\partial u}{\partial x}\right)_{\substack{x=z\\y=0}} = 0, \qquad \phi_2(z,0) = \left(\frac{\partial u}{\partial y}\right)_{\substack{x=z\\y=0}} = -3z^2$$

$$f'(z) = \phi_1(z,0) - i\phi_2(z,0)$$

$$= 0 + i3z^2$$

Integrating we get

$$f(z) = iz^3 + c \qquad \text{where } c \text{ is a complex constant.}$$

Example 13: Prove that the function $u = e^x(x\cos y - y\sin y)$ satisfies Laplace equation and find the corresponding analytic function.

Solution: Given, $u = e^x(x\cos y - y\sin y)$

$$\therefore \quad \frac{\partial u}{\partial x} = e^x(x\cos y - y\sin y) + e^x\cos y$$

$$\frac{\partial u}{\partial y} = e^x(-x\sin y - \sin y - y\cos y)$$

and $$\frac{\partial^2 u}{\partial x^2} = e^x(x\cos y - y\sin y) + e^x\cos y + e^x(\cos y) \qquad (1)$$

$$\frac{\partial^2 u}{\partial y^2} = e^x(-x\cos y - \cos y - \cos y + y\sin y) \qquad (2)$$

Adding (1) & (2) we get

$$\therefore \quad \frac{\partial^2 u}{\partial x^2} + \frac{\partial^2 u}{\partial y^2} = 0$$

Since u satisfies the Laplace equation and hence it is a harmonic function.

Now, $\phi_1(z,0) = \left(\frac{\partial u}{\partial x}\right)_{\substack{x=z\\y=0}} = e^z(z+1) \qquad \phi_2(z,0) = \left(\frac{\partial u}{\partial y}\right)_{\substack{x=z\\y=0}} = 0$

By Milne's method

$$f'(z) = \phi_1(z,0) - i\phi_2(z,0)$$
$$= e^z(z+1) - i0$$

Integrating

$$f(z) = \int e^z(z+1)dz + c$$
$$= ze^z + c$$

$\therefore \quad f(z) = ze^z + c$ where c is a complex constant.

Example 14: Find an analytic function whose real part is

$$u = \frac{\sin 2x}{\cos h\, 2y - \cos 2x}$$

Solution: Given $u = \dfrac{\sin 2x}{\cos h\, 2y - \cos 2x}$

$$\therefore \quad \frac{\partial u}{\partial x} = \frac{(\cos h\, 2y - \cos 2x)(2\cos 2x) - (\sin 2x)(2\sin 2x)}{(\cos h\, 2y - \cos 2x)^2}$$

$$= \frac{2(\cos 2x \cos h\, 2y - 1)}{(\cos h\, 2y - \cos 2x)^2} = \phi_1(x,y) \qquad (1)$$

and $$\frac{\partial u}{\partial y} = -\frac{2(\sin 2x)(\sin h\, 2y)}{(\cos h\, 2y - \cos 2x)^2} = \phi_2(x, y) \tag{2}$$

By Milne's method

$$f'(z) = \phi_1(z,0) - i\phi_2(z,0)$$

$$= \frac{2(\cos 2z - 1)}{(1 - \cos 2z)^2} - i(0) \qquad \text{using (1) and (2)}$$

$$= \frac{-2}{(1 - \cos 2z)} = \frac{-2}{2\sin^2 z} = -\operatorname{cosec}^2 z$$

Integrating

$$f(z) = \int -\operatorname{cosec}^2 z\, dz + c$$

$$= \cot z + c$$

$$\therefore \quad f(z) = \cot z + c, \quad \text{where } c \text{ is a complex constant.}$$

Example 15: If $f(z) = u + iv$ where u and v are functions of x and y is an analytic function show that

$$\left[\frac{\partial}{\partial x}|f(z)|\right]^2 + \left[\frac{\partial}{\partial y}|f(z)|\right]^2 = |f'(z)|^2$$

Solution: Given $f(z) = u + iv$ be an analytic function so that

$$|f(z)|^2 = u^2 + v^2$$

$$\therefore \quad |f(z)| = \sqrt{u^2 + v^2}$$

Now,

$$\frac{\partial}{\partial x}|f(z)| = \frac{1}{2\sqrt{u^2 + v^2}}\left(2u\frac{\partial u}{\partial x} + 2v\frac{\partial v}{\partial x}\right)$$

and $$\frac{\partial}{\partial y}|f(z)| = \frac{1}{2\sqrt{u^2 + v^2}}\left(2u\frac{\partial u}{\partial y} + 2v\frac{\partial v}{\partial y}\right)$$

$$= \frac{1}{2\sqrt{u^2 + v^2}}\left(-u\frac{\partial v}{\partial x} + v\frac{\partial u}{\partial x}\right) \quad \text{by } C\text{–}R \text{ equations}$$

Squaring and adding, we get

$$\left|\frac{\partial}{\partial x}|f(z)|\right|^2 + \left|\frac{\partial}{\partial y}|f(z)|\right|^2 = \frac{u^2+v^2}{u^2+v^2}\left[\left(\frac{\partial u}{\partial x}\right)^2 + \left(\frac{\partial v}{\partial x}\right)^2\right]$$

$$= \left(\frac{\partial u}{\partial x}\right)^2 + \left(\frac{\partial v}{\partial x}\right)^2$$

$$= |f'(z)|^2$$

Example 16: If $w = f(z)$ is a regular function of z. Prove that

$$\left(\frac{\partial^2}{\partial x^2}\right) + \left(\frac{\partial^2}{\partial y^2}\right) \log|f'(z)| = 0$$

Proof. We have

$$\frac{\partial^2}{\partial x^2} + \frac{\partial^2}{\partial y^2} = 4\frac{\partial^2}{\partial z \partial z}$$

$$\therefore \quad \left(\frac{\partial^2}{\partial x^2} + \frac{\partial^2}{\partial y^2}\right)\{\log|f'(z)|\} = 4\frac{\partial^2}{\partial z \partial z}\left\{\frac{1}{2}\log|f'(z)|^2\right\}$$

$$= 2\frac{\partial^2}{\partial z \partial \bar{z}}\{\log f'(z) \cdot f'(\bar{z})\}$$

$$= 2\frac{\partial^2}{\partial z \partial \bar{z}}\{\log f'(z) + \log f'(\bar{z})\}$$

$$= 2\frac{\partial^2}{\partial \bar{z}}\left\{\frac{1}{f'(z)}f''(z)\right\}$$

$$= 0$$

Since $f'(z)$ and $f''(z)$ are independent.

Example 17: Find an analytic function $f(z) = u + iv$ given that

$$u = x^2 - y^2 + \frac{x}{x^2+y^2}$$

Solution: Let $f(z) = u + iv$ be an analytic function

$$u = x^2 - y^2 + \frac{x}{x^2+y^2}$$

$$\therefore \quad \frac{\partial u}{\partial x} = 2x + \frac{(x^2+y^2) - x \times 2x}{(x^2+y^2)^2}$$

$$= 2x + \frac{y^2 - x^2}{(x^2+y^2)^2} = \phi_1(x,y) \qquad (1)$$

$$\text{and} \quad \frac{\partial u}{\partial y} = -2y - \frac{2xy}{(x^2+y^2)^2} = \phi_2(x,y) \qquad (2)$$

By Milne's method

$$f^{\text{I}}(z) = \phi_1(z,0) - i\phi_2(z,0)$$

$$= \left(2z - \frac{1}{z^2}\right) - i(0).$$

Integrating we get

$$f(z) = \int \left(2z - \frac{1}{z^2}\right) dz + c$$

$$f(z) = z^2 + \frac{1}{z} + c, \quad \text{where } c \text{ is a complex constant.}$$

Example 18: Find an analytic function $f(z) = u + iv$ given that

$$u - v = (x-y)(x^2+4xy+y^2)$$

Solution: Now,

$$\frac{\partial}{\partial x}(u-v) = \frac{\partial u}{\partial x} - \frac{\partial v}{\partial x} = (x^2+4xy+y^2) + (x-y)(2x+4y)$$

$$= 3x^2 + 6xy - 3y^2 \qquad (1)$$

$$\text{and} \quad \frac{\partial}{\partial y}(u-v) = \frac{\partial u}{\partial y} - \frac{\partial v}{\partial y} = -(x^2+4xy+y^2) + (x-y)(4x+2y)$$

$$= 3x^2 - 6xy - 3y^2 \qquad (2)$$

Using C–R equations $\frac{\partial u}{\partial x} = \frac{\partial v}{\partial y}$ and $\frac{\partial u}{\partial y} = -\frac{\partial v}{\partial x}$ and adding (1) and (2)

$$2\frac{\partial u}{\partial y} = 6x^2 - 6y^2 \qquad (3)$$

Again using C–R equations and subtracting (2) from (1)

$$2\frac{\partial u}{\partial x} = 12xy \qquad (4)$$

Now,

$$f'(z) = \frac{\partial u}{\partial x} + i\frac{\partial v}{\partial x}$$

$$= \frac{\partial u}{\partial x} - i\frac{\partial u}{\partial y}$$

$$= 6xy - i(3x^2 - 3y^2) \qquad \text{using (3) and (4)}$$

$$= -3i(x^2 - y^2 + 2ixy)$$

$$= -3i(x + iy)^2$$

$$= -3i\, z^2$$

Integrating, we get

$$f(z) = -i\frac{3}{3}z^3 + c$$

$$= -iz^3 + c \quad \text{where } c \text{ is a complex constant.}$$

Example 19: Find an analytic function $f(z) = u + iv$ given that $u - v = e^x(\cos y - \sin y)$.

Solution: Let $f(z) = u + iv$ be an analytic function.

Now,

$$\frac{\partial}{\partial x}(u - v) = \frac{\partial u}{\partial x} - \frac{\partial v}{\partial x} = e^x(\cos y - \sin y)$$

$$\text{and} \quad \frac{\partial}{\partial y}(u - v) = \frac{\partial u}{\partial y} - \frac{\partial v}{\partial y} = e^x(-\sin y - \cos y)$$

By using *C–R* equations

$$\frac{\partial v}{\partial y} - \frac{\partial v}{\partial x} = e^x(\cos y - \sin y) \qquad (1)$$

$$\text{and} \quad \frac{\partial v}{\partial x} + \frac{\partial v}{\partial y} = e^x(\sin y + \cos y) \qquad (2)$$

Adding (1) and (2) we get

$$\frac{\partial v}{\partial y} = e^x \cos y$$

and subtracting (2) from (1) we get

$$\frac{\partial v}{\partial x} = e^x \sin y$$

Consider

$$f^{\text{I}}(z) = \frac{\partial v}{\partial y} + i\frac{\partial v}{\partial x}$$

$$= e^x \cos y + ie^x \sin y$$

$$= e^x e^{iy} = e^{x+iy} = e^z$$

Integrating,

$$f(z) = e^z + c, \quad \text{where } c \text{ is a complex constant.}$$

Example 20: Show that the function $u = \frac{1}{2}\log(x^2 + y^2)$ is harmonic and find its harmonic conjugate.

Solution: Given, $u = \frac{1}{2}\log(x^2 + y^2)$

$$\therefore \quad \frac{\partial u}{\partial x} = \frac{x}{x^2 + y^2}; \qquad \frac{\partial u}{\partial y} = \frac{y}{x^2 + y^2}$$

$$\text{and} \quad \frac{\partial^2 u}{\partial x^2} = \frac{y^2 - x^2}{(x^2 + y^2)^2}; \qquad \frac{\partial^2 u}{\partial y^2} = \frac{x^2 - y^2}{(x^2 + y^2)^2}$$

$$\text{Adding,} \quad \frac{\partial^2 u}{\partial x^2} + \frac{\partial^2 u}{\partial y^2} = 0.$$

Since u satisfies the Laplace equation, it is a harmonic function.

Let v be the conjugate harmonic function to u. Then

$$dv = \frac{\partial v}{\partial x}dx + \frac{\partial v}{\partial y}dy$$

$$= -\frac{\partial u}{\partial y}dx + \frac{\partial u}{\partial x}dy \qquad \text{by } C\text{–}R \text{ equations}$$

$$= \frac{-y}{x^2 + y^2}dx + \frac{x}{x^2 + y^2}dy$$

$$= \frac{xdy - ydx}{x^2 + y^2}$$

Integrating, we get

$$v = \tan^{-1}\frac{y}{x} + c \text{ where } c \text{ is a complex constant.}$$

Example 21: If $w = \phi + i\psi$ represents the complex potential for an electric field and $\psi = x^2 - y^2 + \dfrac{x}{x^2 + y^2}$, determine the function ϕ.

Solution: Given, $\psi = x^2 - y^2 + \dfrac{x}{x^2 + y^2}$

$$\therefore \quad \frac{\partial \psi}{\partial x} = 2x + \frac{y^2 - x^2}{(x^2 + y^2)^2}$$

$$\frac{\partial \psi}{\partial y} = -2y + \frac{-2xy}{(x^2 + y^2)^2}$$

and
$$\frac{\partial^2 \psi}{\partial x^2} = 2 + \frac{(x^2 + y^2)^2(-2x) - (y^2 - x^2)2(x^2 + y^2)2x}{(x^2 + y^2)^4}$$

$$\frac{\partial^2 \psi}{\partial y^2} = -2 + \frac{(x^2 + y^2)^2(-2) - (-2y)(2)(x^2 + y^2)2y}{(x^2 + y^2)^4}$$

Adding, we get

$$\frac{\partial^2 \psi}{\partial x^2} + \frac{\partial^2 \psi}{\partial y^2} = 0.$$

$\therefore$ ψ satisfies the Laplace equation and hence it is harmonic.

Now, by *C–R* equations

$$\frac{\partial \phi}{\partial x} = \frac{\partial \psi}{\partial y} = \frac{\partial}{\partial y}\left[x^2 - y^2 + \frac{x}{x^2 + y^2}\right] = -2y - \frac{2xy}{(x^2 + y^2)^2}$$

Integrating w.r.t. x

$$\phi = \int\left(-2y - \frac{2xy}{(x^2 + y^2)^2}\right)dx + c$$

$$\therefore \quad \phi = -2xy + \frac{y}{x^2 + y^2} + c_1(y) \quad \text{where } c(y) \text{ is a function of } y.$$

where

$$c_1(y) = c$$

$$\therefore \quad \phi = -2xy + \frac{y}{x^2 + y^2} + c.$$

Example 22: Determine the analytic function $f(z) = u + iv$ if

$$u - v = \frac{\cos x + \sin x - e^{-y}}{2(\cos x - \cosh y)} \text{ and } f\left(\frac{\pi}{2}\right) = 0.$$

Solution: Given

$$u - v = \frac{\cos x + \sin x - e^{-y}}{2(\cos x - \cosh y)}$$

$$\therefore \quad \frac{\partial u}{\partial x} - \frac{\partial v}{\partial x} = \frac{\partial}{\partial x}(u - v) = \frac{(\sin x - \cos x)\cosh y + 1 - e^{-y}\sin x}{2(\cos x - \cosh y)^2} \tag{1}$$

$$\text{and } \frac{\partial u}{\partial y} - \frac{\partial v}{\partial y} = \frac{\partial}{\partial y}(u - v) = \frac{(\cos x - \cosh y)e^{-y} + (\cos x + \sin x - e^{-y})\sinh y}{2(\cos x - \cosh y)^2}$$

$$\text{or} \quad -\frac{\partial v}{\partial x} - \frac{\partial u}{\partial x} = \frac{(\sin x + \cos x)\sinh y + e^{-y}(\cos x - \cosh y - \sinh y)}{2(\cos x - \cosh y)^2} \tag{2}$$

Adding (1) and (2)

$$-2\frac{\partial v}{\partial x} = \frac{(\sin x - \cos x)\cosh y + (\sin x + \cos x)\sinh y + 1 + e^{-y}(-\sin x + \cos x - \cosh y - \sinh y)}{2(\cos x - \cosh y)^2}$$

Subtracting (2) from (1) and

$$2\frac{\partial u}{\partial x} = \frac{(\sin x - \cos x)\cosh y - (\sin x + \cos x)\sinh y + 1 - e^{-y}(\sin x + \cos x - \cosh y - \sinh y)}{2(\cos x - \cosh y)^2}$$

Now

$$f'(z) = \frac{\partial u}{\partial x} + i\frac{\partial v}{\partial x}$$

$$= \frac{(1 - \cos z)}{2(1 - \cos z)^2} \qquad \text{(Putting } x = z,\ y = 0\text{)}$$

$$= \frac{1}{2(1 - \cos z)}$$

$$= \frac{1}{4\sin^2 \frac{z}{2}} = \frac{1}{4}\operatorname{cosec}^2 \frac{z}{2}$$

Integrating, we get

$$f(z) = -\frac{1}{2}\cot\frac{z}{2} + c$$

Since $f\left(\frac{\pi}{2}\right) = 0,\ 0 = -\frac{1}{2}\cot\frac{\pi}{4} + c \quad \therefore \quad c = \frac{1}{2}$

$$\therefore \quad f(z) = \frac{1}{2}\left(1 - \cot\frac{z}{2}\right)$$

QUESTION BANK

1. Show that the function $f(z) = \sqrt{|xy|}$ is not analytic at the origin even though C–R equations are satisfied thereat.

2. Determine the analytic function whose real part is $u = x^3 - 3xy^2 + 3x^2 - 3y^2 + 1$

   Ans: $z^3 + 3z^2 + 1 + ic$

3. Determine the analytic function whose real part is $u = e^{2x}(x\cos 2y - y\sin 2y)$

   Ans: $ze^{2z} + ic$

4. Find the analytic function whose real part is $u = x\sin x\cos h\, y - y\cos x \sin h\, y$

   Ans: $z\sin z$

5. Find the regular function whose imaginary part is $v = e^{-x}(x\cos y + y\sin y)$

   Ans: $1 + ize^{-z}$

6. Find the regular function whose imaginary part is $v = \dfrac{2\sin x\sin y}{\cos 2x + \cos h\, 2y}$

   Ans: $(2\cos x\cos h\, y)/(\cos 2x + \cos h\, 2y)$

7. If $f(z) = u + iv$ is an analytic function of z then find $f(z)$ if $u - v = (x - y)(x^2 + 4xy + y^2)$

   Ans: $c - iz^3$

8. An electostatic field in the xy-plane is given by the potential function $\phi = 3x^2y - y^3$. Find the stream function.

   Ans: $\psi = 3xy^2 - x^3 + c$

9. If the potential function is $\log(x^2 + y^2)$, find the flux function and the complex potential function.

   Ans: $2\tan^{-1}\left(\frac{y}{x}\right)$

10. In a two-dimensional fluid flow, the stream function ψ is given, find the velocity potential ϕ whose $\psi = \tan^{-1}\left(\frac{y}{x}\right)$.

Ans: $\frac{1}{2}\log(x^2 + y^2)$

EXAMINATION QUESTIONS

1. Define an analytic function and obtain Cauchy-Reimann equations in cartesian form.

2. If $f(z) = u(r, \theta) + iv(r, \theta)$ is analytic show that

$$\frac{\partial u}{\partial r} = \frac{1}{r}\frac{\partial v}{\partial \theta}; \quad \frac{\partial v}{\partial r} = -\frac{1}{r}\frac{\partial u}{\partial \theta}$$

3. Define harmonic function. If $f(z) = u + iv$ is an analytic function then show that u and v are both harmonic.

4. Find the analytic function $f(z)$ whose imaginary part is $v = e^x(x\sin y + y\cos y)$.

5. Find the analytic function $f(z) = u + iv$ where

$$u - v = (x - y)(x^2 + 4xy + y^2)$$

6. If $u = \frac{x^2}{y}$, $y \neq 0$ and $v = x^2 + 2y^2$. Show that the curves u = constant and v = constant are orthogonal but $f(z) = u + iv$ is not an analytic function.

7. If $f(z) = u + iv$ where u and v are functions of x, y is an analytic function show that

$$\left[\frac{\partial |f(z)|}{\partial x}\right]^2 + \left[\frac{\partial |f(z)|}{\partial y}\right]^2 = |f'(z)|^2$$

8. Find the analytic functions whose real part is

$$u = x\sin x\cosh y - y\cos x\sinh y$$

9. Find the analytic function $f(z) = u + iv$. Given

$$u = e^{-x}\{(x^2 - y^2)\cos y + 2xy\sin y\}$$

10. If $f(z) = u(x,y) + iv(x,y)$ is analytic show that

$$\left[\frac{\partial^2}{\partial x^2} + \frac{\partial^2}{\partial y^2}\right] |f(z)|^2 = 4|f'(z)|^2$$

11. Find an analytic function $f(z) = u + iv$ given that

$$u = x^2 - y^2 + \frac{x}{x^2 + y^2}$$

12. Show that $u = e^{2x}(x\cos 2y - y\sin 2y)$ is harmonic. Find the analytic function $f(z) = u + iv$.

13. Determine the analytic function whose real part is $u = y + e^x \cos y$.

14. Show that $u = x^3 - 3xy^2 + 3x^2 - 3y^2$ is harmonic function and determine the corresponding analytic function.

15. If $f(z) = u(x,y) + iv(x,y)$ is analytic in an open set S, then obtain the C–R equations

$$\frac{\partial u}{\partial x} = \frac{\partial v}{\partial y}, \quad \frac{\partial u}{\partial y} = -\frac{\partial v}{\partial x}.$$

UNIT 4

Complex Variables – II

4.1 GEOMETRICAL REPRESENTATION OF *w* = *f(z)*

Let us consider the complex function

$$w = f(z)$$

$$\text{i.e.,} \quad u + iv = f(x + iy) \tag{1}$$

This function involves four real variables x, y, u and v. Hence, a four-dimensional region is required to plot (1) in the cartesian fashion. As it is not possible to have 4-dimensional graph paper, we make use of two complex planes one for the variable $z = x + iy$ and the other for the variable $w = u + iv$. If the point z describes some curve c in the z-plane, the point w will move along a corresponding curve c' in the w-plane. Since to each point (x, y) there corresponds a point (u, v), we then say that:

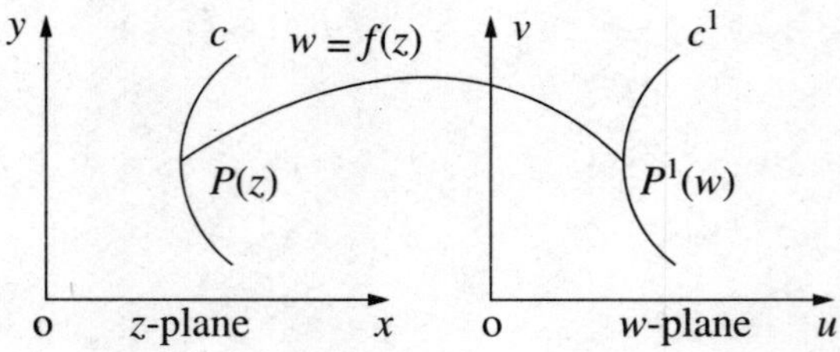

"A curve c in the z-plane is mapped into the corresponding curve c' in the w-plane by the function $w = f(z)$ which defines a mapping or the transformation of the z-plane into the w-plane".

4.1.1 Conformal Transformation

Let us suppose that the two curves c, c_1 intersect at the point p in the z-plane and the corresponding curves c' and c_1' intersect at p' in the w-plane. If the angle of intersection of curves at p is the same as the angle of intersection of the curves at p' in magnitude and sense then the transformation is said to be conformal.

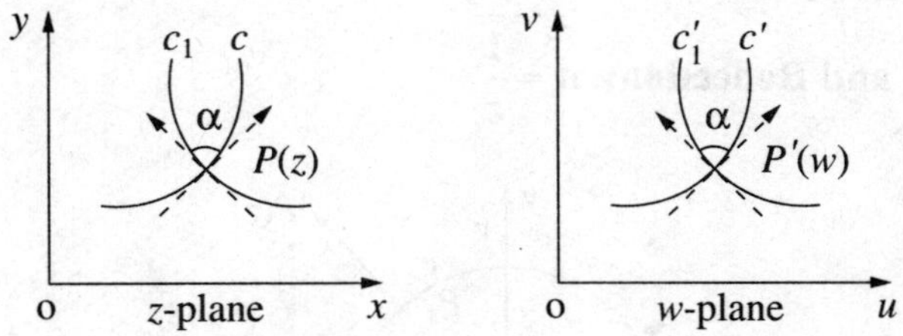

4.1.2 Some Standard Transformations

1. Translation: $w = z + c$ where c is a complex constant.

If $z = x + iy$, $c = c_1 + ic_2$ and $w = u + iv$ then the transformation $w = z + c$ becomes

$$u + iv = x + iy + c_1 + ic_2$$

$$= (x + c_1) + i(y + c_2)$$

$$\therefore \quad u = x + c_1, \; v = y + c_2$$

i.e., the point $p(x,y)$ in the z-plane is mapped onto the point $p'(x + c_1, y + c_2)$ in the w-plane. Every point in the z-plane is mapped onto the w-plane in the same way.

The transformation $w = z + c$ maps a figure in the z-plane onto a figure in the w-plane of the same shape and size. In particular this transformation changes circles into circles.

2. Magnification and Rotation: $w = cz$ where c is a complex constant.

Let $c = \delta e^{i\alpha}$, $z = re^{i\theta}$ and $w = Re^{i\phi}$ be the polar forms.

Then $w = cz$

$$Re^{i\phi} = \delta e^{i\alpha} \cdot r\, e^{i\theta}$$

$$= \delta r\, e^{i(\theta + \alpha)}$$

$$\therefore \quad R = \delta r, \; \phi = \theta + \alpha$$

i.e., the point $p(r, \theta)$ in the z-plane is mapped onto the point $p'(\delta\alpha, \theta + \alpha)$ in the w-plane. Hence, this transformation consists of magnification (or contraction) of the radius vector of p by $\delta = |c|$ and its rotation through an angle

$\alpha = amp(c)$. Accordingly any figure in the z-plane is transformed into a geometrically similar figure in the w-plane. In particular this transformation maps circles onto circles.

3. Inversion and Reflection: $w = \dfrac{1}{z}$

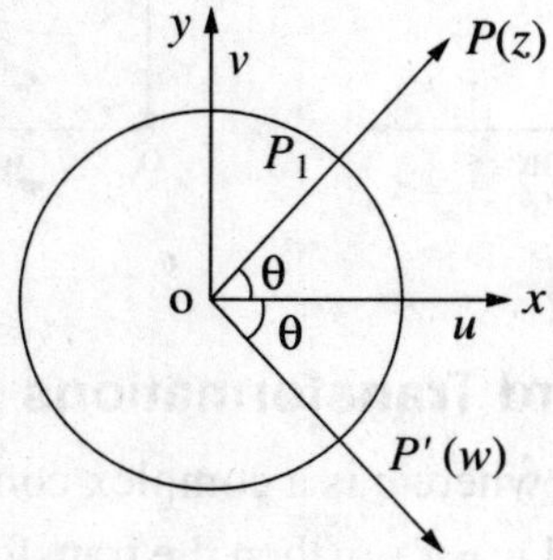

Let us suppose that w-plane is superposed on z-plane. If $z = re^{i\theta}$ and $w = Re^{i\phi}$ then $Re^{i\phi} = \dfrac{1}{re^{i\theta}} = \dfrac{1}{r}e^{-i\theta}$

$$\therefore \quad R = \frac{1}{r} \quad \text{and} \quad \phi = -\theta.$$

Thus, if P be (r, θ) and P_1 be $\left(\dfrac{1}{r}, \theta\right)$ i.e., P_1 is the inverse of P w.r.t. the unit circle with centre O, then the reflection P' and P_1 in the real axis represents $w = \dfrac{1}{z}$. Hence, the transformation is inversion of z w.r.t. the unit circle $|z| = 1$ follows by reflection of the inverse into the real axis.

Question: Prove that the transformation $w = \dfrac{1}{z}$ maps a circle onto a circle or a straight line if the former goes through the origin.

Proof. To prove this we write $z = \dfrac{1}{w}$.

$$\therefore \quad x + iy = \frac{1}{u + iv}$$

$$= \frac{u - iv}{u^2 + v^2}$$

$$\therefore \quad x = \frac{u}{u^2 + v^2} \tag{1}$$

and

$$y = \frac{-v}{u^2 + v^2} \tag{2}$$

Consider the general equation of any circle in the z-plane as

$$x^2 + y^2 + 2gx + 2fy + c = 0 \tag{3}$$

Substituting (1) and (2) in (3)

$$\frac{u^2}{(u^2+v^2)^2} + \frac{(-v)^2}{(u^2+v^2)^2} + 2g\frac{u}{u^2+v^2} + 2f\frac{(-v)}{u^2+v^2} + c = 0$$

$$\therefore \quad c(u^2 + v^2) + 2gu - 2fv + 1 = 0 \tag{4}$$

This is the equation of the circle in the w-plane.

If $c = 0$ the circle (3) passes through the origin and its image (4) reduces to

$$2gu - 2fv + 1 = 0 \tag{5}$$

Which is the equation of a straight line in the w-plane.

4.1.3 Bilinear Transformation

The transformation

$$w = \frac{az + b}{cz + d} \tag{1}$$

where a, b, c and d are complex constants and $ad - bc \neq 0$, is known as bilinear transformation. The condition $ad - bc \neq 0$ ensures that $\frac{dw}{dz} \neq 0$, i.e., the transformation is conformal.

If $ad - bc = 0$ then every point of the z-plane is a critical point.

The inverse mapping of (1) is

$$z = \frac{-dw + b}{cw + a} \tag{2}$$

which is also bilinear transformation.

The general bilinear transformation transforms circles into circles.

Example 1: Find the bilinear transformation which maps the points $z = 1$, i, -1 into the points $w = i,\ 0,\ -i$. Hence, find the image of $|z| < 1$.

Solution: Let

$$w = \frac{az+b}{cz+d} \tag{1}$$

be the linear transformation.

Given $z = 1 \quad i \quad -1$

$w = i \quad 0 \quad -i$

Substituting the three pairs of values of z and w in (1) we get

$$i = \frac{a+b}{c+d}, \quad o = \frac{ai+b}{ci+d}, \quad -i = \frac{-a+b}{-c+d}$$

$$\therefore \quad a+b-i(c+d) = 0 \tag{2}$$

$$ai + b = 0 \tag{3}$$

$$-a+b+i(-c+d) = 0 \tag{4}$$

From (3) we get $b = -ai$

Adding (2) and (4)

$$2b + i(-2c) = 0$$

$$\therefore \quad c = \frac{b}{i} = \frac{-ai}{i} = -a$$

Subtracting (4) from (2)

$$2a - 2id = 0$$

$$\therefore \quad d = -ai$$

Substituting for b, c, d in (1) we get

$$w = \frac{az-ai}{-az-ai} = \frac{z-i}{-z-i}$$

$$\therefore \quad w = \frac{i-z}{i+z} \tag{5}$$

which is the required bilinear transformation.

To find the image of $|z| < 1$ we write (5) as

$$z = i\frac{1-w}{1+w}$$

Since $|z| < 1$ we have

$$\frac{|i||1-w|}{|1+w|} = |z| < 1$$

$$\therefore \quad |1-w| < |1+w|$$

$$\therefore \quad |1-(u+iv)| < |1+u+iv|$$

$$\therefore \quad (1-u)^2 + v^2 < (1+u)^2 + v^2$$

$$\text{i.e.,} \quad 1 - 2u + u^2 + v^2 < 1 + 2u + u^2 + v^2$$

$$\therefore \quad 4u > 0, \quad \text{i.e.,} \quad u > 0$$

Hence, the interior of the circle $x^2 + y^2 = 1$ in the z-plane is transformed to the entire half of the w-plane to the right of the imaginary axis.

Example 2: Find the bilinear transformation that maps the points $z = 1, i, -1$ into the points $w = 0, 1, \infty$.

Solution: Let

$$w = \frac{az+b}{cz+d} \tag{1}$$

be the general bilinear transformation.

Given $z = 1, \quad i, \quad -1$

$w = 0, \quad 1, \quad \infty$

corresponding to $(z, w) = (1, 0)$

$$0 = \frac{a+b}{c+d}$$

$$\therefore \quad a + b = 0$$

$$\therefore \quad b = -a \tag{2}$$

when $(z, w) = (-1, \infty)$ we write equation (1) as

$$\frac{1}{w} = \frac{cz+d}{az+b}$$

$$\therefore \quad 0 = \frac{-c+d}{-a+b} \qquad \left(w \to \infty, \frac{1}{w} \to 0\right)$$

$$\therefore \quad -c + d = 0$$

$$\therefore \quad d = c \tag{3}$$

Corresponding to $(z, w) = (i, 1)$

$$1 = \frac{ai + b}{ci + b}$$

$$\therefore \quad ci + d = ai + b$$

Using (2) and (3) we get

$$ci + c = ai + (-a)$$

$$\therefore \quad c = a\frac{(i-1)}{(i+1)} = d \tag{4}$$

substituting b, c, d in (1)

$$w = \frac{az + (-a)}{a\left(\frac{i-1}{i+1}\right)z + a\left(\frac{i-1}{i+1}\right)}$$

$$= \frac{i+1}{i-1}\,\frac{z-1}{z+1}$$

$$= \frac{i+1}{-1+i} \times \frac{-1-i}{-1-i}\,\frac{z-1}{z+1}$$

$$w = -i\,\frac{z-1}{z+1} \tag{5}$$

which is the required bilinear transformation.

1. Transformation of $w = z^2$

Given $w = z^2$

$$\text{i.e.,} \qquad u + iv = (x + iy)^2$$

$$= x^2 - y^2 + i\,2xy$$

$$\therefore \quad u = x^2 - y^2, \quad v = 2xy$$

Case I: If $u = a$ (constant) then $x^2 - y^2 = a$ which is a rectangular hyperbola.

If $v = b$ (constant) then $xy = \frac{b}{2}$ which also represents a rectangular hyperbola.

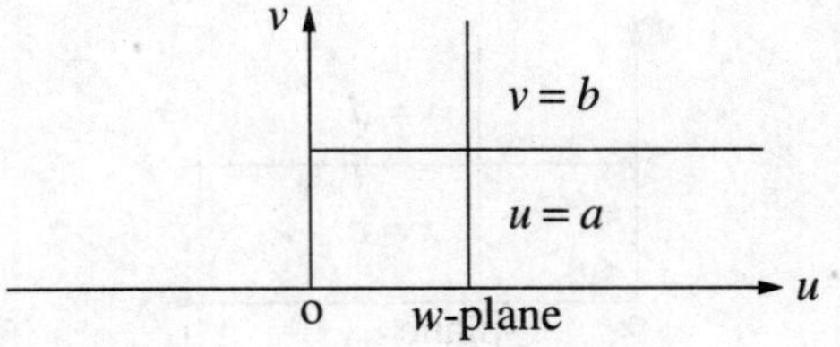

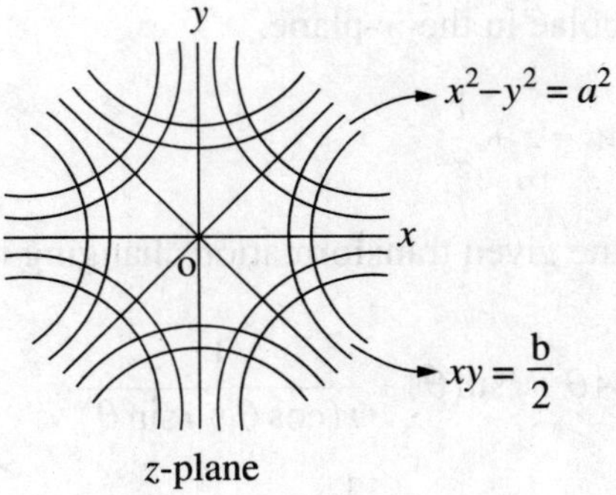

Hence, a pair of lines $u = a,\ v = b$ parallel to the axes in the w-plane map into pair of orthogonal hyperbolae in the z-plane.

Case II: If $x = c$ (constant) then $y = \dfrac{v}{2c}$ and $y^2 = c^2 - u$.

Eliminating y from these equations we get

$v^2 = 4c^2(c^2 - u)$ which represents a parabola.

And if $y = d$ (constant) then $x = \dfrac{v}{2c}$ and $x^2 = u + d^2$.

Eliminating x from these equations we get

$v^2 = 4d^2(d^2 + u)$ which is also a parabola.

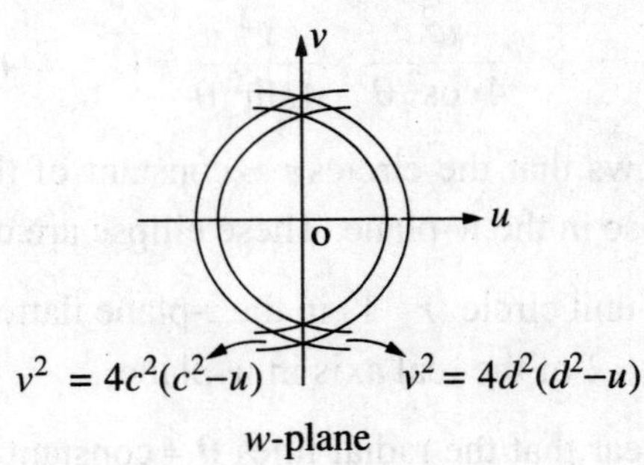

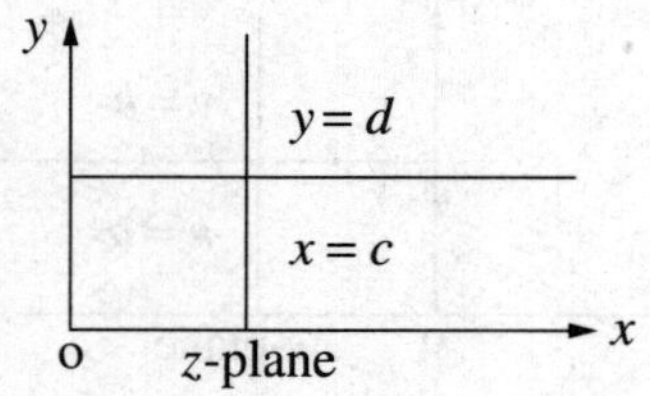

Hence, a pair of lines $x = c$ and $y = d$ parallel to the axes in the z-plane map into orthogonal parabolae in the w-plane.

2. Transformation $w = z + \frac{1}{z}$

Let $w = z + \frac{1}{z}$ be the given transformation changing to polar co-ordinates.

$$u + iv = r(\cos\theta + i\sin\theta) + \frac{1}{r(\cos\theta + i\sin\theta)}$$

$$= r(\cos\theta + i\sin\theta) + \frac{1}{r}(\cos\theta - i\sin\theta) \qquad \text{Taking conjugate}$$

$$= \left(r + \frac{1}{r}\right)\cos\theta + i\left(r - \frac{1}{r}\right)\sin\theta$$

$$\therefore \quad u = \left(r + \frac{1}{r}\right)\cos\theta, \quad v = \left(r - \frac{1}{r}\right)\sin\theta$$

Eliminating θ we get

$$\frac{u^2}{\left(r + \frac{1}{r}\right)^2} + \frac{v^2}{\left(r - \frac{1}{r}\right)^2} = 1 \tag{1}$$

And eliminating r, we get

$$\frac{u^2}{4\cos^2\theta} - \frac{v^2}{4\sin^2\theta} = 1 \tag{2}$$

From (1) it follows that the circles r = constant of the z-plane transform into a family of ellipse in the w-plane. These ellipse are confocal.

In particular, the unit circle $(r = 1)$ in the z-plane flattens out to become the segment $u = -2$ to $u = 2$ of the real axis in w-plane.

From (2), it is clear that the radial lines θ = constant of the z-plane transform into a family of hyperbolae which are confocal.

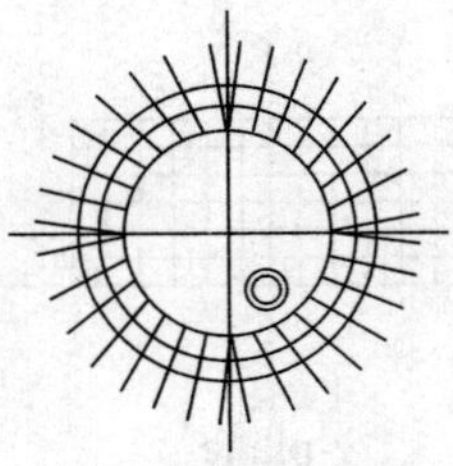

z-plane

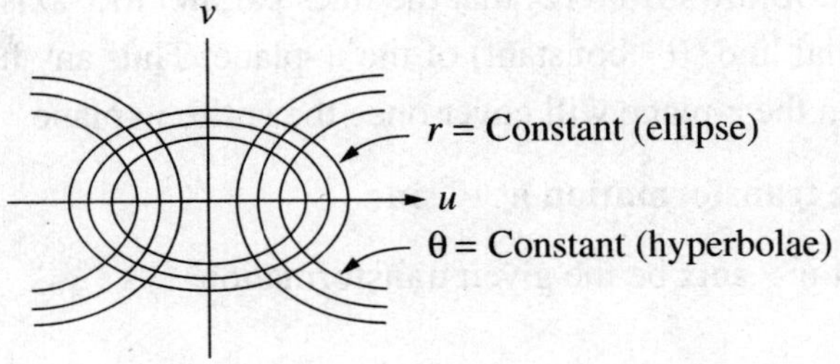

w-plane

3. Transformation of $w = e^z$

Let $z = x + iy$ and $w = \rho e^{i\theta}$ then

$$\rho e^{i\phi} = e^{x+iy} = e^x . e^{iy}$$

$$\therefore \quad \rho = e^x \tag{1}$$

and

$$\phi = y \tag{2}$$

From (1) it is clear that the lines parallel to *y*-axis (x = constant) map into circles (ρ = constant) in the *w*-plane their radii being $\gtrless 1$ according as $x \gtrless 0$

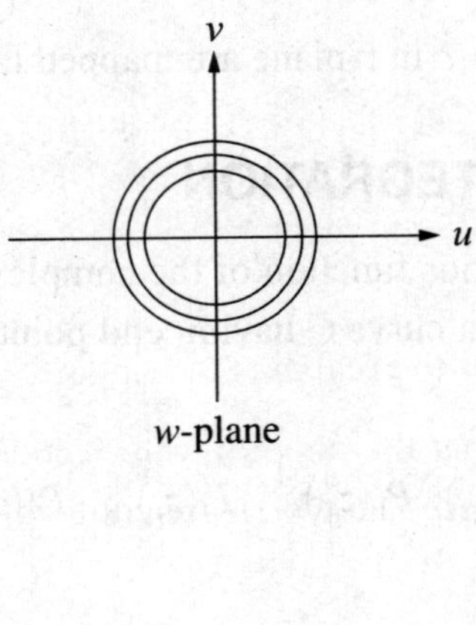

w-plane

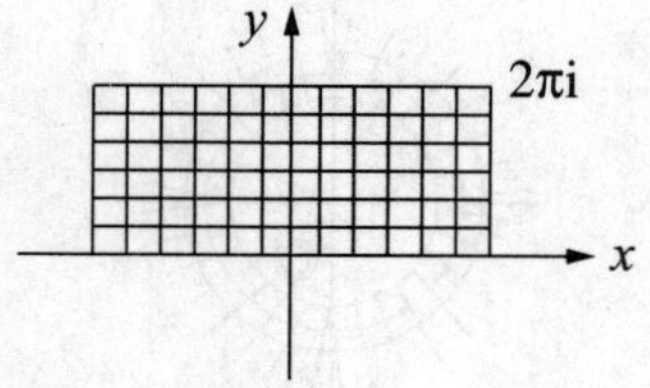

z-plane

Similarly, it follows from (2) that the lines parallel to x-axis (y = constant) map into a radial line (θ = constant) of the w-plane. Thus any horizontal strip of height $2\pi i$ in the z-plane will cover once the entire w-plane.

4. Discuss the transformation $w = \sin z$

Solution: Let $w = \sin z$ be the given transformation

$$\therefore \quad u + iv = \sin(x + iy)$$

$$= \sin x \cosh y + i \cos x \sinh y$$

$$\therefore \quad u = \sin x \cosh y, \quad v = \cos x \sinh y \qquad \textbf{(1)}$$

Eliminating y from equation (1), we get

$$\frac{u^2}{\sin^2 x} - \frac{v^2}{\cos^2 x} = 1 \qquad (\because \quad \cosh^2 y - \sinh^2 y = 1)$$

Thus, the straight lines $x = c$ in the z-plane are mapped into confocal hyperbolas in the w-plane.

Eliminating x from equations (1) we get

$$\frac{u^2}{\cosh^2 y} + \frac{v^2}{\sinh^2 y} = 1$$

Thus, the straight lines $y = c$ in z-plane are mapped into confocal ellipses.

4.2 COMPLEX INTEGRATION

Let $w = f(z)$ be a continuous function of the complex variable $z = (x + iy)$ and is defined at all points of a curve C having end points A and B. Divide C into n parts at the points

$$A = P_0(z_0),\ P_1(z_1), \ldots P_i(z_i), \ldots P_n(z_n) = B.$$

Let $\delta z_i = z_i - z_{i-1}$ and ξ_i be any point on the arc $P_{i-1} - P_i$. The limit of the sum $\sum_{i=1}^{n} f(\xi_i)\delta z_i$ as $n \to \infty$ in such a way that the length of the chord δz_i approaches zero, is called the line integral of $f(z)$ taken along the path C, i.e., $\int_C f(z)dz$.

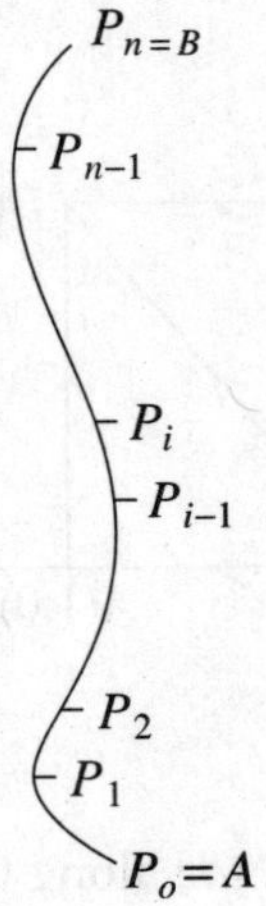

Writing $f(z) = u(x,y) + iv(xy)$ and noting that $dz = dx + idy$

$$\int_C f(z)dz = \int_C (udx - vdy) + i\int_C (vdx + udy)$$

which shows that the evaluation of the line integral of a complex function can be reduced to the evaluation of two line integrals of real functions.

Example 1: Evaluate $\int_0^{i+1} (x - y + ix^2)dz$ along the straight line from $z = 0$ to $z = 1 + i$.

Solution: Along the straight line OP joining $O(z = 0)$ and $P(z = 1 + i)$, $y = x$, $dy = dx$ and x varies from $x = 0$ to $x = 1$

$$\therefore \quad \int_0^{i+1} (x - y + ix^2)dz$$

$$= \int_0^{i+1} (x - y + ix^2)(dx + idy)$$

$$= \int_0^1 (x - x + ix^2)(dx + idx) = \int_0^1 ix^2(1+i)dx$$

$$= (-1+i)\int_0^1 x^2 dx = (-1+i)\left(\frac{1}{3}\right) = -\frac{1}{3} + \frac{1}{3}i$$

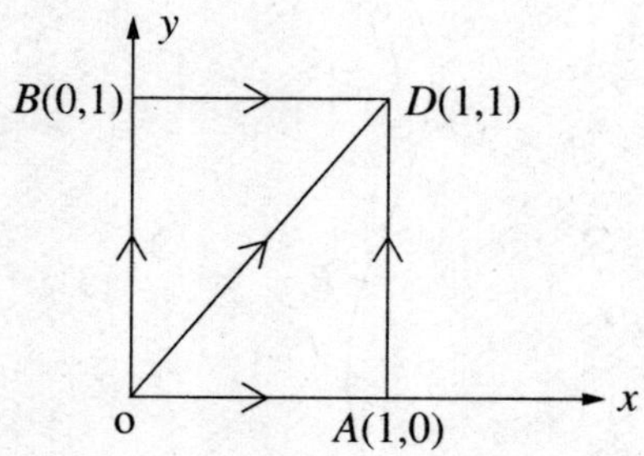

Example 2: Evaluate $\int_0^{1+i} (x^2 - iy)dz$ along the paths

(a) $y = x$ and (b) $y = x^2$

Solution: (a) Along the line $y = x$

$$dy = dx$$

So that $dz = dx + idx$

$$= (1+i)dx$$

$$\therefore \int_0^{1+i} (x^2 + iy)dz$$

$$= \int_0^1 (x^2 - ix)(1+i)dx$$

$$= (1+i)\left[\frac{x^3}{3} - i\frac{x^2}{2}\right]_0^1 = (1+i)\left(\frac{1}{3} - \frac{1}{2}i\right)$$

$$= \frac{(1+i)(2-3i)}{6} = \frac{5}{6} - \frac{1}{6}i$$

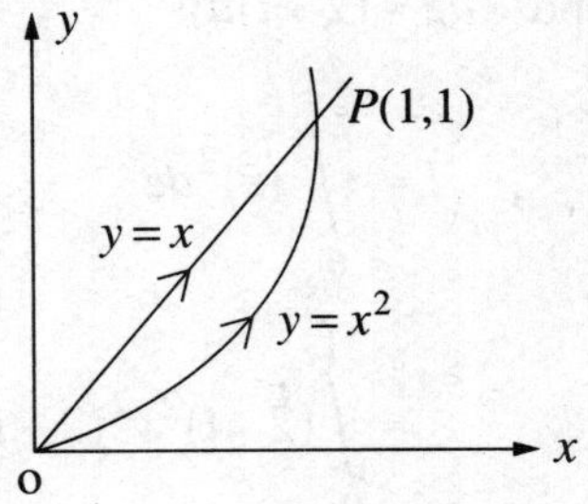

(b) Along the parabola $y = x^2$, $dy = 2x\,dx$ so that

$dz = dx + 2ix\,dx = (1 + 2ix)dx$

and x varies from 0 to 1

$$\therefore \quad \int_0^{1+i} (x^2 - iy)dz = \int_0^1 (x^2 - ix^2)(1 + 2ix)dx$$

$$= (1 - i)\int_0^1 x^2(1 + 2ix)dx$$

$$= (1 - i)\left[\frac{x^3}{3} + i\frac{x^4}{2}\right]_0^1$$

$$= (1 - i)\left(\frac{1}{3} + \frac{1}{2}i\right)$$

$$= \frac{(1 - i)(2 + 3i)}{6} = \frac{5}{6} + \frac{1}{6}i$$

Example 3: Evaluate $\int_0^{2+i} (\bar{z})^2 dz$ along (i) the line $y = \dfrac{x}{2}$

(ii) the real axis to 2 and then vertically to $2 + iy$.

Solution: (i) Along the line OA,

$$x = 2y$$

$$\therefore \quad z = (2 + i)y$$

$$\bar{z} = (2 - i)y$$

and $\quad dz = (2+i)dy$

$$\therefore \quad I = \int_0^{2+i} (\bar{z})^2 dz$$

$$= \int_0^1 (2-i)^2 y^2 (2+i) dy$$

$$= 5(2-i)\left[\frac{y^3}{3}\right]_0^1 = \frac{5}{3}(2-i)$$

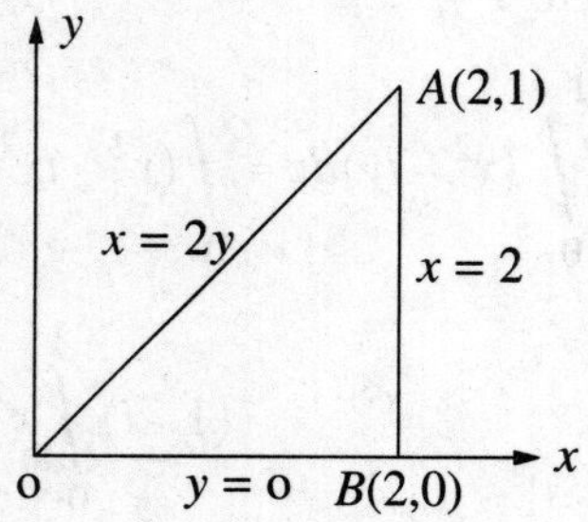

(ii) The real axis to 2 and then vertically to $2+iy$.

$$I = \int_{OB} (\bar{z})^2 dz + \int_{BA} (\bar{z})^2 dz$$

Now, along OB, $z = x$, $\bar{z} = x$ and $dz = dx$

Along BA, $z = 2+iy$, $\bar{z} = 2-iy$, $dz = idy$.

$$\therefore \quad I = \int_0^2 x^2 dx + \int_0^1 (2-iy)^2 idy$$

$$= \left[\frac{x^3}{3}\right]_0^2 + \int_0^1 [4y + (4-y^2)i]dy$$

$$= \frac{8}{3} + 4\cdot\frac{1}{2} + \left(4\cdot 1 + \frac{1}{3}\right)i$$

$$= \frac{1}{3}(14+11i)$$

Example 4: Prove that

(i) $\int_c \frac{dz}{z-a} = 2\pi i$ (ii) $\int_c (z-a)^n dz = 0$, [n, an integer $\neq -1$]

where c is the circle $|z-a| = r$.

Solution: The parametric equation of c is $z - a = re^{i\theta}$ where θ varies from 0 to 2π as z describes c once in the positive sense.

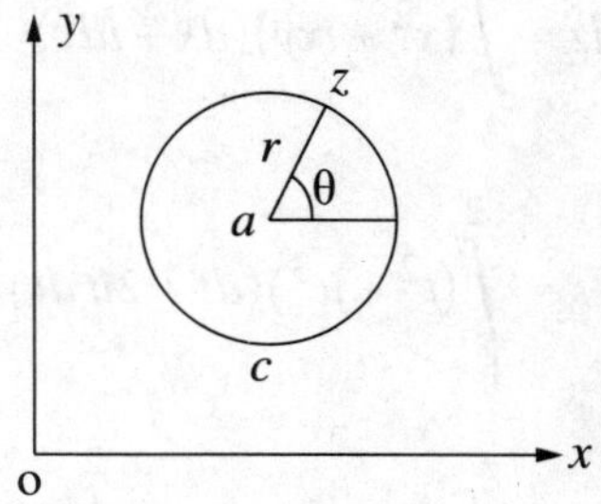

(i)
$$\int_c \frac{dz}{z-a} = \int_0^{2\pi} \frac{1}{re^{i\theta}} ire^{i\theta} d\theta \quad [\because \quad dz = ire^{i\theta} d\theta]$$

$$= i\int_0^{2\pi} d\theta = i[\theta]_0^{2\pi}$$

$$= 2\pi i$$

(ii) $$\int_c (z-a)^n dz$$

$$= \int_0^{2\pi} r^n e^{in\theta} ire^{i\theta} d\theta = ir^{n+1} \int_0^{2\pi} e^{(n+1)i\theta} d\theta$$

$$= ir^{n+1} \left[\frac{e^{(n+1)i\theta}}{i(n+1)}\right]_0^{2\pi} = \frac{r^{n+1}}{n+1}\left[e^{2(n+1)\pi i} - 1\right] = 0 \quad [\because e^{2(n+1)\pi i} = 1]$$

$$\therefore \quad \int_c (z-a)^n dz = 0.$$

Example 5: Integrate $f(z) = x^2 + ixy$ from $A(1,1)$ to $B(2,4)$ along the curve $x = t,\ y = t^2$.

Solution: Equations of the path of integration are $x = t,\ y = t^2$

$$\therefore \quad dx = dt,\ dy = 2t\,dt$$

At $A(1,1)$, $t = 1$ and at $B(2,4)$, $t = 2$.

$$\therefore \quad \int_A^B f(z)dz = \int_A^B (x^2 + ixy)(dx + idy)$$

$$= \int_1^2 (t^2 + it^3)(dt + 2it\,dt)$$

$$= \int_1^2 (t^2 + 2t^4)dt + i\int_1^2 3t^2 dt$$

$$= \left[\frac{t^3}{3} - \frac{2t^5}{5}\right]_1^2 + i\left[\frac{3t^4}{4}\right]_1^2$$

$$= \left(\frac{8}{3} - \frac{64}{5}\right) - \left(\frac{1}{3} - \frac{2}{5}\right) + i\left(12 - \frac{3}{4}\right)$$

$$= \frac{151}{5} + \frac{45}{4}i$$

4.2.1 Simply and Multiple Connected Regions

A curve is called simple closed curve if it does not cross itself. A curve which crosses itself is called a multiple curve.

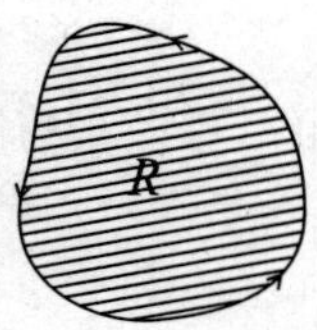

Simple closed curve

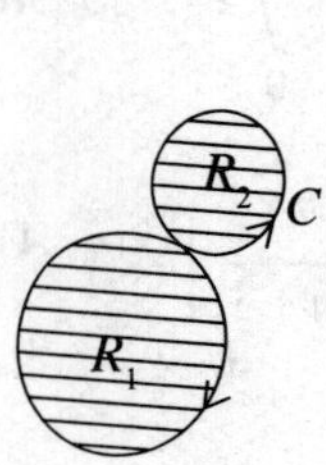

Multiple curve

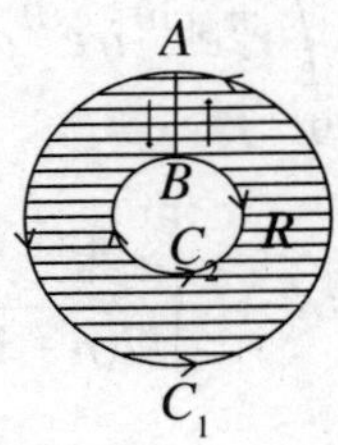

Multiple connected region

A region is called simply connected if every closed curve in the region encloses points of the region only, i.e., every closed curve lying in it can be contracted indefinitely without passing out of it. A region which is not simply connected is called a multiply connected region. In plain terms, a simply connected region is one which has no holes. Figure (3) shows a multiply connected region R enclosed between two separate curves c_1 and c_2. We can convert a multiply connected region into a simply connected one, by giving it one or more cuts (i.e., along the line AB in (3)).

4.3 CAUCHY'S INTEGRAL THEOREM

Statement: If $f(z)$ is an analytic function and $f'(z)$ is continuous at each point within and on a single closed curve C, then

$$\int_c f(z)dz = 0$$

Proof. Let $f(z) = u + iv$ be an analytic function and R be the region bounded by the curve c.

Now,

$$\int_c f(z)dz = \int_c (u+iv)(dx+idy) \quad \text{(where } dz = dx + idy\text{)}$$

$$= \int_c (udx - vdy) + i\int (vdx + udy) \qquad (1)$$

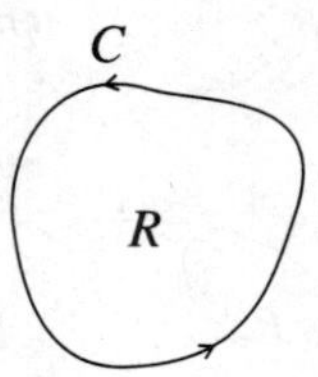

Since $f'(z)$ is continuous, the partial derivatives $\frac{\partial u}{\partial x}, \frac{\partial u}{\partial y}, \frac{\partial v}{\partial x}, \frac{\partial v}{\partial y}$ are also continuous in R. Hence, by Green's theorem we have

$$\int_c f(z)dz = \iint_R \left(-\frac{\partial v}{\partial x} - \frac{\partial u}{\partial y}\right) dxdy + i\iint_R \left(\frac{\partial u}{\partial x} - \frac{\partial v}{\partial y}\right) dxdy \qquad (2)$$

Now $f(z)$ being analytic at each point of the region R by $C - R$ equations we have

$$\frac{\partial u}{\partial x} = \frac{\partial v}{\partial y}, \quad \text{and} \quad \frac{\partial u}{\partial y} = -\frac{\partial v}{\partial x}$$

Thus, the integrals of the two double integrals in (2) vanish identically.

Hence, $\int_c f(z)dz = 0.$

Cor1: If $f(z)$ is analytic in a region R and A and B are two points in R, then $\int_A^B f(z)dz$ is independent of the path joining A and B and lies entirely in R.

Proof. Let APB and AQB be any two paths joining A and B.

By Cauchy's theorem

$$\int_{APBQA} f(z)dz = 0$$

i.e., $$\int_{APB} f(z)dz - \int_{AQB} f(z)dz = 0$$

Hence, $$\int_{APB} f(z)dz = \int_{AQB} f(z)dz$$

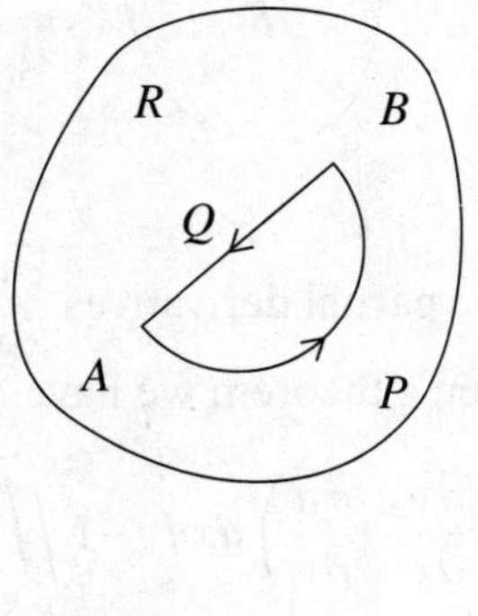

4.3.1 Extension of Cauchy's Theorem

If $f(z)$ is an analytic function in the region bounded by two simple closed curves c and c_1 then

$$\int_c f(z)dz = \int_{c_1} f(z)dz$$

Proof. Let $f(z)$ be an analytic function in the given region.

Let AB be a cross-cut joining the curves c and c_1, then the doubly connected region becomes simply connected.

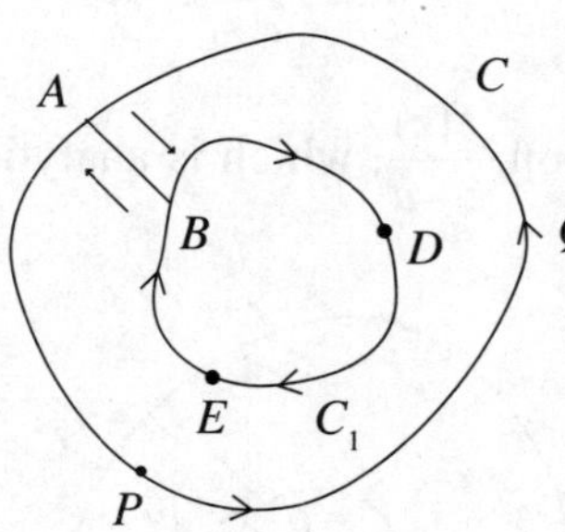

By Cauchy's theorem

$$\int f(z)dz = 0$$

$$\Rightarrow \int_{APQA} f(z)dz + \int_{AB} f(z)dz + \int_{BDEB} f(z)dz + \int_{BA} f(z)dz = 0$$

$$\Rightarrow \int_c f(z)dz + \int_{AB} f(z)dz - \int_{c_1} f(z)dz - \int_{AB} f(z)dz = 0$$

$$\Rightarrow \int_c f(z)dz - \int_{c_1} f(z)dz = 0$$

$$\therefore \int_c f(z)dz = \int_{c_1} f(z)dz$$

(Each integration being taken in anticlockwise direction).

Further, if $c_1, c_2, c_3, \ldots$ be any number of closed curves within c then

$$\int_c f(z)dz = \int_{c_1} f(z)dz + \int_{c_2} f(z)dz + \ldots$$

4.4 CAUCHY'S INTEGRAL FORMULA

Statement: If $f(z)$ is analytic within and on a closed curve c and a is any point within c, then

$$f(a) = \frac{1}{2\pi i}\int_c \frac{f(z)dz}{z-a}$$

Proof. Consider the function $\frac{f(z)}{z-a}$, which is analytic at every point within c except at $z = a$.

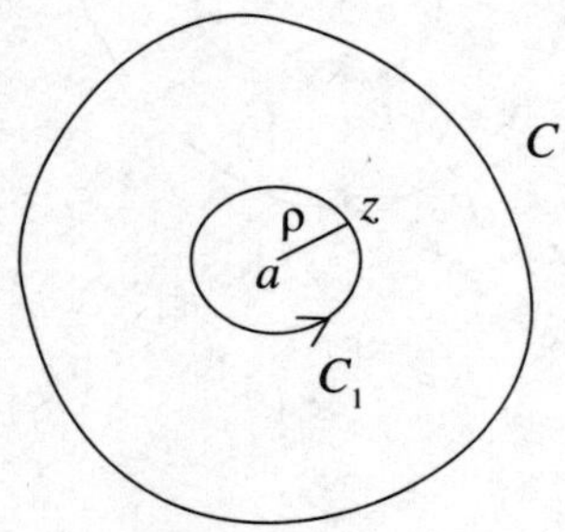

Draw a circle c_1 within 'a' as its centre and radius ρ such that C_1 lies entirely inside C.

Then $\frac{f(z)}{z-a}$ is analytic in the region between C and C_1.

$\therefore$ By Cauchy's theorem, we have

$$\int_C \frac{f(z)}{z-a}dz = \int_{C_1} \frac{f(z)}{z-a}dz \tag{1}$$

Now, equation of circle C_1 is

$$|z-a| = \rho \quad \text{or} \quad z-a = \rho e^{i\theta}$$

$$\therefore \quad dz = i\rho e^{i\theta} d\theta$$

$$\int_{c_1} \frac{f(z)}{z-a}dz = \int_0^{2\pi} \frac{f(a+\rho e^{i\theta})}{\rho e^{i\theta}} i\rho e^{i\theta} d\theta$$

$$= i\int_0^{2\pi} f(a+\rho e^{i\theta})d\theta \qquad (2)$$

In the limit form, as the circle C_1 shrinks to the point a, i.e., as $\rho \to 0$ then the integral (2) will approach to

$$= i\int_0^{2\pi} f(a)d\theta = if(a)\int_0^{2\pi} d\theta$$

$$= 2\pi i f(a)$$

Thus,

$$\int_C \frac{f(z)}{z-a}dz = 2\pi i f(a)$$

$$\therefore \quad f(a) = \frac{1}{2\pi i}\int_C \frac{f(z)}{z-a}dz \qquad (3)$$

Which is the desired Cauchy's integral formula.

Cor: Differentiating (3) both sides w.r.t. 'a'

$$f'(a) = \frac{1}{2\pi i}\int_C \frac{\partial}{\partial a}\left(\frac{f(z)}{z-a}\right)dz$$

$$= \frac{1}{2\pi i}\int_C \frac{f(z)}{(z-a)^2}dz$$

Similarly,

$$f''(a) = \frac{2!}{2\pi i}\int_C \frac{f(z)}{(z-a)^3}dz$$

and in general

$$f^n(a) = \frac{n!}{2\pi i}\int_C \frac{f(z)}{(z-a)^{n+1}}dz \qquad (4)$$

Thus, it follows from the results (3) and (4) that if a function $f(z)$ is known to be analytic on the simple closed curve C then the values of the function and all its derivatives can be found at any point of C. Incidently, we have established a remarkable fact that an analytic function possesses derivatives of all orders and these are themselves all analytic.

In other words, if a function of a complex variable has the first derivative in a simply connected region, all its higher derivatives exist in that region. This property is not exhibited by the functions of real variables.

Example 1: Evaluate $\int_C \frac{z^2 - z + 1}{z - 1} dz$, where C is a cricle

(i) $|z| = 1$ (ii) $|z| = \frac{1}{2}$

Solution: (i) Here $f(z) = z^2 - z + 1$ and $a = 1$

Since $f(z)$ is analytic within and on circle C

$|z| = 1$ and $a = 1$ lies on c.

By Cauchy's integral formula

$$\frac{1}{2\pi i}\int \frac{f(z)}{z-a} dz = f(a) = 1$$

$$\text{i.e.,} \quad \int \frac{z^2 - z + 1}{z - 1} dz = 2\pi i$$

(ii) In this case $a = 1$ lies outside the circle c

$|z| = \frac{1}{2}$, so $\frac{z^2 - z + 1}{z - 1}$ is analytic.

$\therefore$ By Cauchy's theorem

$$\int_C \frac{z^2 - z + 1}{z - 1} dz = 0.$$

Example 2: Evaluate, using Cauchy's integral formula

(i) $\int_C \frac{e^{2z}}{(z-1)(z-2)} dz$ where c is the circle $|z| = 3$

(ii) $\int_C \frac{\cos \pi z}{z^2 - 1} dz$ around a rectangle with vertices $2 \pm i, -2 \pm i$

Solution: (i) Here $f(z) = e^{2z}$ is analytic within circle $C, |z| = 3$ and the two singular points $a = 1$ and $a = 2$ lies inside C.

$$\therefore \quad \int_C \frac{e^{2z}}{(z-1)(z-2)} dz = \int_C e^{2z}\left(\frac{1}{z-2} - \frac{1}{z-1}\right) dz$$

$$= \int_C \frac{e^{2z}}{z-2} dz - \int_C \frac{e^{2z}}{z-1} dz = 2\pi i e^4 - 2\pi i e^2$$

$$= 2\pi i(e^4 - e^2), \quad \text{by Cauchy's integral formula}$$

(ii) $f(z) = \cos \pi z$ is analytic in the region bounded by the given rectangle and the two singular points $a = 1$ and $a = -1$ lie in this rectangle.

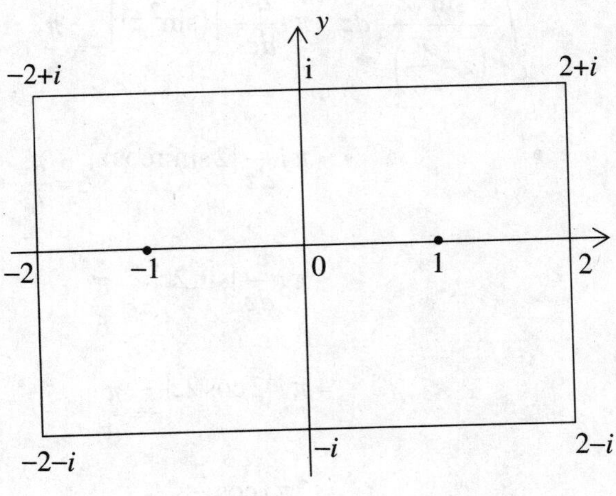

$$\therefore \quad \int_C \frac{\cos \pi z}{z^2 - 1} dz = \frac{1}{2} \int_C \left(\frac{1}{z-1} - \frac{1}{z+1}\right) \cos \pi z \, dz$$

$$= \frac{1}{2}(2\pi i \cos \pi(1)) - \frac{1}{2}(2\pi i \cos \pi(-1))$$

$$= \pi i(\cos \pi - \cos(-\pi))$$

$$= \pi i(-1+1)$$

$$= 0$$

$$\therefore \quad \int_C \frac{\cos \pi z}{z^2 - 1} dz = 0$$

Example 3: Evaluate $\int_C \frac{\sin^2 z}{\left(z - \frac{\pi}{6}\right)^3} dz$ where c is the circle $|z| = 1$.

Solution: Given $f(z) = \sin^2 z$ is analytic inside the circle, $|z| = 1$ and the point $a = \frac{\pi}{6}$ (= 0.5 approx.) lies within c.

$\therefore$ By Cauchy's integral formula

$$f''(a) = \frac{2!}{2\pi i} \int_C \frac{f(z)}{(z-a)^3} dz$$

$$\therefore \quad \int_C \frac{\sin^2 z}{\left(z - \frac{\pi}{6}\right)^3} dz = \pi i \frac{d^2}{dz^2} \left[(\sin^2 z)\right]_{z = \frac{\pi}{6}}$$

$$= \pi i \frac{d}{dz} [2 \sin z \cos z]_{z = \frac{\pi}{6}}$$

$$= \pi i \frac{d}{dz} [\sin 2z]_{z = \frac{\pi}{6}}$$

$$= \pi i [2 \cos 2z]_{z = \frac{\pi}{6}}$$

$$= 2\pi i \cos \frac{\pi}{3}$$

$$= \pi i$$

Example 4: Evaluate $\int_C \frac{e^{2z}}{(z+1)^4} dz$, where c is the circle $|z| = 2$.

Solution: Given, $f(z) = e^{2z}$ is analytic within the circle c, $|z| = 2$

Also $z = -1$ lies inside c.

$\therefore$ By Cauchy's integral formula

$$f'''(a) = \frac{3!}{2\pi i}\int \frac{f(z)}{(z-a)^4}dz$$

$$\therefore \quad \int \frac{e^{2z}}{(z+1)^4}dz = \frac{2\pi i}{6}\left[\frac{d^3}{dz^3}(e^{2z})\right]_{z=-1}$$

$$= \frac{\pi i}{3}\left(8e^{2z}\right)_{z=-1}$$

$$= \frac{8\pi i}{3}e^{-2}$$

Example 5: Evaluate $\int\limits_C \frac{e^z}{(z^2+\pi^2)^2}dz$ where c is the circle $|z| = 4$.

Solution:

Given, $f(z) = \dfrac{e^z}{(z^2+\pi^2)^2}$

$= \dfrac{e^z}{(z+\pi i)^2(z-\pi i)^2}$ is not analytic at $z = \pm\pi i$

However, both $z = \pm\pi i$ lie within the circle $|z| = 4$.

Now

$$\frac{1}{(z+\pi i)^2(z-\pi i)^2} = \frac{A}{(z+\pi i)} + \frac{B}{(z+\pi i)^2} + \frac{C}{(z-\pi i)} + \frac{D}{(z+\pi i)^2}$$

By partial fractions we get

$$A = \frac{7}{2\pi^3 i} \quad C = -\frac{7}{2\pi^3 i} \quad B = D = -\frac{1}{4\pi^2}$$

$$\therefore \quad \int\limits_C \frac{e^z}{(z^2+\pi^2)^2}dz = \frac{7}{2\pi^3 i}\left[\int\limits_C \frac{e^z}{z+\pi i}dz - \int\limits_C \frac{e^z}{z-\pi i}dz\right]$$

$$-\frac{1}{4\pi^2}\left[\int\limits_C \frac{e^z}{(z+\pi i)^2}dz + \int\limits_C \frac{e^z}{(z-\pi i)^2}dz\right]$$

$$= \frac{7}{2\pi^3 i}[2\pi i\, f(-\pi i) - 2\pi i\, f(\pi i)]$$

$$- \frac{1}{4\pi^2}\left[2\pi i\, f'(-\pi i) + 2\pi i\, f'(\pi i)\right]$$

where $f(z) = e^z$

$$= \frac{7}{\pi^2}\left(e^{-\pi i} - e^{\pi i}\right) - \frac{i}{2\pi}\left(e^{-\pi i} + e^{\pi i}\right)$$

$$= -\frac{14i}{\pi^2}\left(\frac{e^{\pi i} - e^{-\pi i}}{2i}\right) - \frac{i}{\pi}\left(\frac{e^{\pi i} + e^{-\pi i}}{2}\right)$$

$$= -\frac{14i}{\pi^2}\sin\pi - \frac{i}{\pi}\cos\pi$$

$$= 0 - \frac{i}{\pi}(-1) \qquad (\because \quad \sin\pi = 0,\ \cos\pi = -1)$$

$$= \frac{i}{\pi}$$

Example 6: If $f(\xi) = \int_C \frac{3z^2 + 7z + 1}{z - \xi}dz$, where c is the circle $x^2 + y^2 = 4$, find the values of $f(3), f'(1 - i)$ and $f''(1 - i)$.

Solution: The given circle c is $x^2 + y^2 = 4$ or $|z| = 2$.

The point $z = 3$ lies outside the circle $|z| = 2$ whereas $z = 1 - i$ and $z = 1 + i$ lie inside the circle.

Now, $f(3) = \int_C \frac{3z^2 + 7z + 1}{z - 3}dz$ and $\frac{3z^2 + 7z + 1}{z - 3}$ is analytic within and on c.

$\therefore$ By Cauchy's integral theorem, we have

$$\int_C \frac{3z^2 + 7z + 1}{z - 3}dz = 0$$

$$\therefore \quad f(3) = 0$$

Now, let $\phi(z) = 3z^2 + 7z + 1$ which is analytic everywhere

$\therefore$ By Cauchy's integral formula, we have

$$\phi(\xi) = \frac{1}{2\pi i}\int \frac{\phi(2)}{z-\xi}\,dz \text{ where } \xi \text{ is a point within } c$$

$$\therefore \quad 2\pi i\phi(\xi) = \int \frac{3z^2+7z+1}{z-\xi}\,dz = f(\xi)$$

$$\therefore \quad f(\xi) = 2\pi i\left(3z^2+7\xi+1\right)$$

$$\therefore \quad f'(1-i) = 2\pi i\,(6(1-i)+7)$$

$$= 2\pi i(13-6i)$$

$$= 2\pi(6+13i)$$

$$\text{Also} \quad f''(\xi) = 12\pi i$$

$$\therefore \quad f''(1-i) = 12\pi i$$

Example 7: If $F\xi = \int\limits_C \frac{4z^2+z+5}{z-\xi}\,dz$, where c is the ellipse $\left(\frac{x}{2}\right)^2 + \left(\frac{4}{2}\right)^2 = 1.$

Find the value of (a) $F(3.5)$ (b) $F(i)$ $F'(-1)$ and $F''(-i)$

Solution: (a) Given $F(\xi) = \int\limits_C \frac{4z^2+z+5}{z-\xi}\,dz$

$$\therefore \quad F(3.5) = \int\limits_C \frac{4z^2+z+5}{z-\xi}\,dz$$

Since $\xi = 3.5$ is the only singular point of $\frac{4z^2+z+5}{z-3.5}$ and it lies outside the ellipse c. Therefore $\frac{4z^2+z+5}{(z-3.5)}$ is analytic everywhere within c. Hence by Cauchy's theorem

$$\int\limits_C \frac{4z^2+z+5}{z-3.5}\,dz = 0 \quad \text{i.e., } F(3.5) = 0$$

(b) Since $f(z) = 4z^2 + z + 5$ is analytic within c and $\xi = i,\ -1$ and $-i$ all lie within c

$\therefore$ by Cauchy's integral formula

$$f(\xi) = \frac{1}{2\pi i}\int\limits_C \frac{f(z)}{z-\xi}\,dz$$

$$\text{i.e.,} \int_C \frac{4z^2+z+5}{z-\xi}dz = 2\pi i\left(4\xi^2+\xi+5\right)$$

$$\therefore \quad F(\xi) = 2\pi i(4\xi^2+\xi+5)$$

$$F'(\xi) = 2\pi i(8\xi+1) \text{ and } F''(\xi) = 16\pi i$$

Thus,

$$F(i) = 2\pi i(-4i^2 - i + 5)$$

$$= 2\pi i(i+1)$$

$$F'(-1) = 2\pi i(-8+1)$$

$$= -14\pi i$$

$$\alpha F''(-i) = 16\pi i$$

QUESTION BANK

1. Discuss the transformation $w = e^z$.
2. Find the bilinear transformation which maps the points $1, 1, -1$ onto the points $2, 1, -2$ respectively.
3. Find the bilinear transformation that maps $z = 1, i, -1$ onto the points $w = 0, 1, \infty$.
4. Find the image of the unit circle in the z – plane under the transformation $w = z + \frac{1}{z}$.
5. Find the bilinear transformation that maps the points $0, -i, -1$ of z-plane onto the points $i, 1, 0$ of w-plane respectively.
6. Find the bilinear transformation which maps the points $0, 1, i$ in the z-plane onto $1+i, -i, 2-i$ in the w-plane.
7. What is the bilinear transformation which maps the points $z = -1, 0, 1$ into the points $w = 0, i, 3i$ respectively.
8. Find the bilinear transformation which maps the points $z = 1, i, -1$ into $w = 2, i, -2$ respectively.

9. Evaluate $\int \frac{3z^2 + 7z + 1}{z + 1} dz$ where c is the circle

(a) $|z| = 1.5$ (b) $|z + i| = 1$

Ans: (a) $-6\pi i$ (b) 0

10. Evaluate $\int_C \frac{z^3 + z + 1}{z^2 - 7z + 2} dz$, where c is the ellipse $4x^2 + 94^2 = 1$

Ans: 0

11. Evaluate $\int_C \frac{\cos \pi z}{z^2 - 1} dz$ around the rectangle with vertices

(a) $z \pm i, \; -z \pm i$ b) $-i, \; 2 - i, \; 2 + i, \; i$

Ans: (a) 0 (b) $-\pi i$

12. Evaluate $\int_C \frac{\sin^2 z}{\left(2 - \frac{\pi}{6}\right)^3} dz$ where c is the circle $|z| = 1$

Ans: πi

13. Evaluate $\int_C \frac{\sin \pi z^2 + \cos \pi z^2}{(z - 1)(z - 2)} dz$ where c is the circle $|z| = 3$.

Ans: $4\pi i$

14. Evaluate $\int_C \frac{e^z}{(z - 1)(z - 4)} dz$, where c is the circle $|z| = 2$

Ans: $-\frac{2}{3} \pi i \, e$

15. Evaluate $\int_C \frac{e^z}{(z + 3)(z + 2)} dz$ where c is the circle $|z - 1| = \frac{1}{2}$.

Ans: 0

16. Evaluate $\int_C \frac{e^{-2z}}{(z + 1)^3} dz$ where c is the circle $|z| = 2$

Ans: $4\pi i$

17. Evaluate $\int_C \frac{dz}{z^2 - 4}$ over (i) $|z| = 1$ and (ii) $|z + 2| = 1$

18. Evaluate $\int_C \tan z \, dz$, where c is $|z| = 2.5$

19. Evaluate using Cauchy's integral formula $\int_C \frac{z}{z^2 - 3z + 2} dz$ where c is the circle $|z - 2| = \frac{1}{2}$.

Ans: $4\pi i$

20. Evaluate $\int_C \frac{z^3 - 2z + 1}{(z - i)^2} dz$ where c is $|z| = 2$

Ans: $-10\pi i$

UNIT 5

Special Functions

5.1 SOLUTION OF LAPLACE EQUATION IN CYLINDRICAL AND SPHERICAL SYSTEMS LEADING TO BESSEL'S AND LEGENDRE'S DIFFERENTIAL EQUATIONS

Laplace's equation has wide applications in physics and engineering. The theory of its solutions is called the *potential theory* and its solutions are called *harmonic functions*. The solution of Laplace's equation, subject to certain boundary conditions, is simplified by a proper choice of coordinate system.

(a) If the problem involves rectangular boundaries, we prefer to take Laplace equation in cartesian coordinates.

The two-dimensional heat flow equation in steady state reduces to

$$\frac{\partial^2 u}{\partial x^2} + \frac{\partial^2 u}{\partial y^2} = 0 \qquad (1)$$

which is Laplace equation in two dimensions.

Also the three-dimensional heat flow equation in steady state reduces to

$$\frac{\partial^2 u}{\partial x^2} + \frac{\partial^2 u}{\partial y^2} + \frac{\partial^2 u}{\partial z^2} = 0 \qquad (2)$$

which is Laplace's equation in three dimensions.

(b) If the problem involves circular boundaries, we prefer to take Laplace's equation in polar coordinates given by

$$\frac{\partial^2 u}{\partial r^2} + \frac{1}{r}\frac{\partial u}{\partial r} + \frac{\partial^2 u}{\partial \theta^2} = 0 \qquad (3)$$

By putting $x = r\cos\theta,\ y = r\sin\theta$ in equation (1) we get equation (3). Thus changing the independent variables from (xy) to $(r,\ \theta)$.

(c) If the problem involves cylindrical boundaries, we prefer to take Laplace's equation in cylindrical coordinates given by

$$\frac{\partial^2 u}{\partial r^2}+\frac{1}{r}\frac{\partial u}{\partial r}+\frac{1}{r^2}\frac{\partial^2 u}{\partial \theta^2}+\frac{\partial^2 u}{\partial z^2}=0 \tag{4}$$

By putting $x = r\cos\theta$, $y = r\sin\theta$, $z = z$ in equation (2) we get equation (4). Thus changing the independent variables (x, y, z) to (r, θ, z).

(d) If the problem involves spherical boundaries, we prefer to take Laplace's equation in spherical coordinates given by

$$\frac{\partial^2 u}{\partial r^2}+\frac{2}{r}\frac{\partial u}{\partial r}+\frac{1}{r^2}\frac{\partial^2 u}{\partial \theta^2}+\frac{\cot\theta}{r^2}\frac{\partial u}{\partial \theta}+\frac{1}{r^2\sin^2\theta}\frac{\partial^2 u}{\partial \phi^2}=0 \tag{5}$$

By putting $x = r\sin\theta\ \cos\phi$, $y = r\sin\theta\sin\phi$, $z = r\cos\theta$ in equation (2) we get equation (5). Thus changing the independent variables (x, y, z) to (r, θ, ϕ).

5.1.1 Solution of Laplace's Equation in Cylindrical Coordinates

Let $$\frac{\partial^2 u}{\partial r^2}+\frac{1}{r}\frac{\partial u}{\partial r}+\frac{1}{r^2}\frac{\partial^2 u}{\partial \theta^2}+\frac{\partial^2 u}{\partial z^2}=0 \tag{1}$$

be the Laplace's equation in cylindrical coordinates.

Let $$u(r, \theta, z) = R(r)S(\theta)T(z) \tag{2}$$

where R is the function of r only, S is the function of θ and T is the function of z alone, be the solution of equation (1)

$$\therefore \quad \frac{\partial^2 u}{\partial r^2}=R''ST,\ \frac{\partial u}{\partial r}=R'ST,\ \frac{\partial^2 u}{\partial \theta^2}=RS''T \quad \text{and} \quad \frac{\partial^2 u}{\partial z^2}=RST''$$

Substituting in (1)

$$R''ST+\frac{1}{r}R'ST+\frac{1}{r^2}RS''T+RST''=0$$

Dividing by RST we get

$$\frac{1}{R}\left(\frac{d^2R}{dr^2}+\frac{1}{r}\frac{dR}{dr}\right)+\frac{1}{r^2 S}\frac{d^2S}{d\theta^2}+\frac{1}{T}\frac{d^2T}{dz^2}=0 \tag{3}$$

Assuming $$\frac{d^2S}{d\theta^2}=-n^2S \text{ and } \frac{d^2T}{dz^2}=k^2T \tag{4}$$

Equation (3) reduces to

$$\frac{1}{R}\left(\frac{d^2R}{dr^2}+\frac{1}{r}\frac{dR}{dr}\right)-\frac{n^2}{r^2}+k^2=0$$

$$\therefore \quad r^2\frac{d^2R}{dr^2}+r\frac{dR}{dr}+\left(k^2r^2-n^2\right)R=0 \tag{5}$$

which is *Bessel's equation*

Its solution is

$$R=C_1J_n(kr)+C_2Y_n(kr)$$

and the solutions of equations

$$\frac{d^2s}{d\theta^2}+n^2s=0 \quad \text{and} \quad \frac{d^2T}{dz^2}-k^2T=0 \quad \text{are respectively}$$

$$S=C_3\cos n\theta+C_4\sin n\theta \quad \text{and} \quad T=C_5e^{kz}+C_6e^{-kz}$$

Hence, the solution of equation (1) is

$$u=RST$$

i.e., $u(r\ \theta\ z)=[C_1J_n(kr)+C_2Y_n(kr)]\,[C_3\cos n\theta+C_4\sin n\theta]\left[C_5e^{kz}+C_6e^{-kz}\right]$

which is known as a cylindrical harmonic.

5.1.2 Solution of Laplace's Equation in Spherical Coordinates

$$\text{Let} \quad \frac{\partial^2u}{\partial r^2}+\frac{z}{r}\frac{\partial u}{\partial r}+\frac{1}{r^2}\frac{\partial^2u}{\partial\theta^2}+\frac{\cot\theta}{r^2}\frac{\partial u}{\partial\theta}+\frac{1}{r^2\sin^2\theta}\frac{\partial^2u}{\partial\phi^2}=0 \tag{1}$$

be the Laplace's equation in spherical coordinates

$$\text{Let} \quad u(r,\theta,\phi)=R(r)\,S(\theta)\,T(\phi) \tag{2}$$

$$\text{or} \quad u=RST$$

where R is the function of r only, S is the function of θ and T is the function of ϕ alone, be the solution of equation (1).

$$\therefore \quad \frac{\partial^2u}{\partial r^2}=R''ST;\ \frac{\partial u}{\partial r}=R'ST,\ \frac{\partial^2u}{\partial\theta^2}=RS''T,\ \frac{\partial u}{\partial\theta}=RS'T,\ \frac{\partial^2u}{\partial\phi^2}=RST''$$

Substituting (1) we get

$$R''ST + \frac{2}{r}R'ST + \frac{1}{r^2}RS''T + \frac{\cot\theta}{r^2}RS'_T + \frac{1}{r^2\sin^2\theta}RST'' = 0$$

Dividing throughout by RST we get

$$\frac{1}{R}\left(r^2\frac{d^2R}{dr^2} + 2r\frac{dR}{dr}\right) + \frac{1}{S}\left(\frac{d^2S}{d\theta^2} + \cot\theta\frac{ds}{d\theta}\right) + \frac{1}{T\sin^2\theta}\frac{d^2T}{d\phi^2} = 0 \quad (3)$$

Putting, $$\frac{1}{R}\left(r^2\frac{d^2R}{dr^2} + 2r\frac{dR}{dr}\right) = n(n+1) \quad (4)$$

and $$\frac{1}{T}\frac{d^2T}{d\phi^2} = -m^2 \quad (5)$$

Equation (3) reduces to

$$n(n+1) + \frac{1}{S}\left(\frac{d^2S}{d\theta^2} + \cot\theta\frac{ds}{d\theta}\right) + \frac{1}{\sin^2\theta}(-m^2) = 0$$

$$\therefore \quad \frac{d^2s}{d\theta^2} + \cot\theta\frac{ds}{d\theta} + \left[n(n+1) - m^2\operatorname{cosec}^2\theta\right]S = 0 \quad (6)$$

Which is *Legendre's equation*

Its solution is

$$S = C_1P_n^m(\cos\theta) + C_2\phi_n^m(\cos\theta)$$

The solution of (5) i.e.

$$\frac{d^2T}{d\phi^2} + m^2T = 0 \quad \text{is}$$

$$T = C_3\cos m\phi + C_4\sin m\phi$$

Now to solve equation (4)

i.e., $$\frac{1}{R}\left(r^2\frac{d^2R}{dr^2} + 2r\frac{dR}{dr}\right) - n(n+1) = 0$$

$$\therefore \quad r^2\frac{d^2R}{dr^2} + 2r\frac{dR}{dr} - n(n+1)R = 0 \quad (1)$$

Put $R = r^k$ so that (1) reduces to

$$k(k-1) + 2k - n(n+1) = 0$$

$$\therefore \quad (k^2 - n^2) + (k - n) = 0$$

$$(k - n)(k + n + 1) = 0 \quad \therefore \quad k = n \quad \text{or} \quad k = -n - 1$$

$$\text{Thus,} \quad R = C_5 r^n + C_6 r^{-n-1}$$

Hence, the general solution of equation (1) is

$$u(r, \theta, \phi) = R(r)S(\theta)T(\phi)$$

$$= \sum_{n=0}^{\infty} \sum_{m=0}^{\infty} [C_1 P_n^m(\cos\theta) + C_2 \phi_n^m(\cos\theta)]$$

$$\times [C_3 \cos m\phi + C_4 \sin m\phi]\left[C_5 r^n + C_6 r^{-n-1}\right]$$

Any solution of (1) is known as a spherical harmonic.

Example 1: Find the potential in the interior of a sphere of unit radius when the potential on the surface is $f(\theta) = \cos^2\theta$.

Solution: The Laplace's equation in spherical form is

$$\frac{\partial^2 u}{\partial r^2} + \frac{2}{r}\frac{\partial u}{\partial r} + \frac{1}{r^2}\frac{\partial^2 u}{\partial \theta^2} + \frac{\cot\theta}{r^2}\frac{\partial u}{\partial \theta} + \frac{1}{r^2 \sin^2\theta}\frac{\partial^2 u}{\partial \phi^2} = 0$$

Take the origin at the centre of the given sphere since the potential is independent of ϕ on sphere, so also is the potential at any point.

$\therefore$ the Laplace's equation in spherical coordinates reduces to

$$\frac{\partial^2 u}{\partial r^2} + \frac{2}{r}\frac{\partial u}{\partial r} + \frac{1}{r^2}\frac{\partial^2 u}{\partial \theta^2} + \frac{\cot\theta}{r^2}\frac{\partial u}{\partial \theta} = 0 \tag{1}$$

Let $u(r, \theta) = R(r)\, S(\theta)$

where R is the function of r only and S is the function of θ alone, be the solution of (1).

$$\therefore \quad \frac{\partial^2 u}{\partial r^2} = R''S, \quad \frac{\partial u}{\partial r} = R'S, \quad \frac{\partial^2 u}{\partial \theta^2} = RS'' \quad \text{and} \quad \frac{\partial u}{\partial \theta} = RS'$$

Substituting these values in equation (1) we get

$$\frac{d^2 s}{d\theta^2} + \cot\theta \frac{ds}{d\theta} + n(n+1)s = 0 \tag{2}$$

$$\text{and} \quad \frac{1}{R}\left(r^2 \frac{d^2 R}{dr^2} + 2r\frac{dR}{dr}\right) = n(n+1) \tag{3}$$

Put $\cot\theta = v$ in equation (2) we get

$$(1-v^2)\frac{d^2s}{dv^2} - 2v\frac{ds}{dv} + n(n+1)s = 0$$

which is *Legendre's equation*

$\therefore$ its solutions are

$$S = P_n(v) = P_n(\cos\theta) \quad \text{for} \quad n = 0, 1, 2 \ldots$$

The solutions of (3) are

$$R_n(r) = r^n,\ \overline{R}_n(r) = r^{-n-1} = \frac{1}{r^{n+1}}$$

Hence, equation (1) has the following two sets of solutions

$$u_n(r,\ \theta) = c_n r^n P_n(\cos\theta) \quad \text{and} \quad \overline{u}_n(r,\ \theta) = c_n P_n \frac{\cos\theta}{r^{n+1}}$$

where $n = 0, 1, 2, 3 \ldots$

For points inside sphere, we have the general equation

$$u(r,\theta) = \sum_{n=0}^{\infty} C_n r^n P_n(\cos\theta) \tag{4}$$

On the boundary of sphere $u(1,\theta) = f(\theta)$.

$$\therefore \quad f(\theta) = \sum_{n=0}^{\infty} C_n P_n(\cos\theta)$$

which is Fourier-Legendre expansion of $f(\theta)$. Hence

$$C_n = \left(n+\frac{1}{2}\right)\int_{-1}^{1} f(\theta)P_n(x)dx \qquad \text{where } x = \cos\theta$$

$$= \left(n+\frac{1}{2}\right)\int_{-1}^{1} x^2 P_n(x)dx \qquad (\because \quad f(\theta) = \cos^2\theta)$$

$$= \left(n+\frac{1}{2}\right)\int_{-1}^{1} \left[\frac{2}{3}P_2(x) + \frac{1}{3}P_0(x)\right] P_n(x)dx \qquad \left[\because \quad P_2(n) = \frac{1}{3}(3x^2-1)\right]$$

Using the orthogonality of Legendre polynomials, we get

$$C_n = 0 \text{ except for } n = 0, 2$$

$$\therefore \quad C_0 = \frac{1}{2} \cdot \frac{1}{3} \int_{-1}^{1} P_0^2(x)dx = \frac{1}{3}$$

$$C_2 = \frac{5}{2} \cdot \frac{2}{3} \int_{-1}^{0} P_2^2(x)dx = \frac{2}{3}$$

Substituting in (4) we get

$$u(r,\theta) = \frac{1}{3} + \frac{2}{3} r^2 P_2(\cos\theta)$$

$$\text{or} \quad u(r,\theta) = \frac{1}{3} r^2 \left(\cos^2\theta - \frac{1}{3}\right)$$

Example 2: Show that in spherical polar coordinates (r, θ, ϕ), Laplace's equation possesses solutions of the form

$$\left(Ar^n + \frac{B}{r^{n+1}}\right) P_n(\mu) e^{\pm im\phi}$$

where $\mu = \cos\theta, A, B, m, n$ are constants and $P_n(\mu)$ satisfies the Legendre's equation

$$(1-\mu^2)\frac{d^2P_n}{d\mu^2} - 2\mu\frac{dP_n}{d\mu} + \left[n(n+1) - \frac{m^2}{1+\mu^2}\right] P_n = 0$$

Ans: Same as in (1)

Example 3: A fluid (wind) with velocity 50 km/hr flows uniformly over a spherical dome having a radius 5 m. The upstream pressure and temperature are equal to those inside the dome. Estimate the upward force on the dome.

Solution: The velocity potential function u, satisfies the Laplace's equation. This equation is spherical coordinates as

$$\frac{1}{r^2}\frac{\partial}{\partial r}\left(r^2\frac{\partial u}{\partial r}\right) + \frac{1}{r^2\sin\theta}\frac{\partial}{\partial\theta}\left(\sin\theta\frac{\partial u}{\partial\theta}\right) + \frac{1}{r^2\sin^2\theta}\frac{\partial^2 u}{\partial\phi^2} = 0 \qquad (1)$$

The velocity components are

$$V_r = \frac{\partial u}{\partial r} \tag{2}$$

$$V_\theta = \frac{1}{r}\frac{\partial u}{\partial \theta} \tag{3}$$

$$\text{and} \quad V_\phi = \frac{1}{r\sin\theta}\frac{\partial u}{\partial \phi} \tag{4}$$

The velocity field satisfies the boundary conditions

$$V_r = 0 \text{ at } r = R \tag{5}$$

$$V_r = V_\infty \sin\theta \sin\phi \tag{6}$$

$$\text{and} \quad V_\phi = V_\infty \cos\phi \tag{7}$$

The solution of the equation by the method of separation of variables,

$u(r,\theta,\phi) = P(r)S(\theta)T(\phi)$, gives three ordinary differential equations.

$$T''(\phi) + m^2 T(\phi) = 0 \tag{8}$$

$$r^2 P''(r) + 2rP'(r) - n(n+1)P(r) = 0 \quad \text{(Euler's differential equation)} \tag{9}$$

and

$$\sin^2\theta S''(\theta) + \sin\theta\cos\theta S'(\theta) + \left[n(n+1)\sin^2\theta - m^2\right] S(\theta) = 0 \tag{10}$$

This is associated Legendre's equation.

The solutions of these equations give

$$u(r,\theta,\phi) = \sum_{n=0}^{\infty}\left(a_n r^n + \frac{b_n}{r^{n+1}}\right)\sum_{m=0}^{n} P_n^m(\cos\theta)\left[A_{nm}\cos m\phi + B_{nm}\sin m\phi\right] \tag{11}$$

where $P_n^m(cos\theta)$ is Legendre polynomial.

By using the boundary conditions and for $m = n = 1$ the velocity potential is obtained as

$$u = V_\infty\left(r + \frac{R^3}{2r^2}\right)\sin\theta\sin\phi$$

5.2 BESSEL'S EQUATION

The differential equation

$$x^2 \frac{d^2y}{dx^2} + x\frac{dy}{dx} + (x^2 - n^2)y = 0 \tag{1}$$

is known as *Bessel's equation of order n*. Its particular solutions are called *Bessel functions of order n*.

Since $x = 0$ is a regular singularity of the equation, let its solution be

$$\begin{aligned} y &= x^m(a_0 + a_1x + a_2x^2 + \ldots) \\ &= a_0x^m + a_1x^{m+1} + a_2x^{m+2} + \ldots \end{aligned} \tag{2}$$

Then,

$$\frac{dy}{dx} = ma_0x^{m-1} + (m+1)a_1x^m + (m+2)a_2x^{m+1} + \ldots$$

$$\frac{d^2y}{dx^2} = m(m-1)a_0x^{m-2} + (m+1)ma_1x^{m-1} + (m+2)(m+1)a_2x^m + \ldots$$

Substituting y, $\frac{dy}{dx}$ and $\frac{d^2y}{dx^2}$ in (1) we get

$$\begin{aligned} & x^2\left[m(m-1)a_0x^{m-2} + (m+1)ma_1x^{m-1} + (m+2)(m+1)a_2x^m + \ldots\right] \\ & + x\left[ma_0x^{m-1} + (m+1)a_1x^m + (m+2)a_2x^{m+1} + \ldots\right] \\ & + (x^2 - n^2)\left[a_0x^m + a_1x^{m+1} + a_2x^{m+2} + \ldots\right] = 0 \end{aligned}$$

$$\therefore \quad a_0(m^2 - n^2)x^m + a_1\left[(m+1)^2 - n^2\right]x^{m+1} + a_2\left[((m+2)^2 - n^2) + a_0\right] x^{m+2} + \ldots = 0$$

Equating to zero the coefficient of x^m, we obtain the *indicial equation* $m^2 - n^2 = 0 \quad (\because \quad a_0 \neq 0)$

$$\therefore \quad (m-n)(m+n) = 0 \quad \therefore \quad m = n \text{ or } -n$$

Equating to zero the coefficient of next term i.e., x^{m+1} we get

$$a_1\left[(m+1)^2 - n^2\right] = 0$$

$$\Rightarrow \quad a_1 = 0 \quad \text{since } (m+1)^2 - n^2 \neq 0 \text{ for } m = \pm n$$

Equating to zero the coefficient of next term i.e., x^{m+2} we get

$$a_2\left[(m+2)^2-n^2\right]+a_0=0$$

$$\therefore \quad a_2=\frac{-a_0}{(m+2)^2-n^2}$$

Continuing in this way we get

$$a_3=0, \quad a_4=\frac{-a_2}{(m+4)^2-n^2} \quad \text{etc.}$$

Thus, $a_1=a_3=a_5=\ldots=0$ and

$$a_2=\frac{-a_0}{(m+2)^2-n^2}, \quad a_4=\frac{-a_2}{(m+4)^2-n^2}$$

$$=\frac{a_0}{[(m+2)^2-n^2][(m+4)^2-n^2]}$$

$$a_6=-\frac{a_0}{[(m+2)^2-n^2][(m+4)^2-n^2][(m+6)^2-n^2]} \quad \text{etc.}$$

Substituting these values in (2) we get

$$y=a_0x^m\left[1-\frac{1}{(m+2)^2-n^2}x^2+\frac{1}{[(m+2)^2-n^2][(m+4)^2-n^2]}x^4\right.$$

$$\left.-\frac{1}{[(m+2)^2-n^2][(m+4)^2-n^2][(m+6)^2-n^2]}x^6+\ldots\right]$$

For $m=n$

$$y_1=a_0x^n\left\{1-\frac{1}{4(n+1)}x^2+\frac{1}{4^2 2!(n+1)(n+2)}x^4\right.$$

$$\left.-\frac{1}{4^3 3!(n+1)(n+2)(n+3)}x^6+\ldots\right\} \qquad (3)$$

and for $m=-n$

$$y_2=a_0x^{-n}\left\{1-\frac{1}{4(-n+1)}x^2+\frac{1}{4^2 2!(-n+1)(-n+2)}x^4\right.$$

$$\left.-\frac{1}{4^3 3!(-n+1)(-n+2)(-n+3)}x^6+\ldots\right\} \qquad (4)$$

Depending upon the value of n, we get different types of solutions.

Case I: When n is not integral or zero, the complete solution of equation (1) is $y = c_1 y_1 + c_2 y_2$.

Take $a_0 = \dfrac{1}{2^n \overline{|n+1}}$ then the solution given in (3) is called the Bessel function of the first kind of order n and is denoted by $J_n(x)$.

Thus

$$J_n(x) = \left(\frac{x}{2}\right)^n \left\{\frac{1}{\overline{|n+1}} - \frac{1}{1!\overline{|n+2}}\left(\frac{x}{2}\right)^2 + \frac{1}{2!\overline{|n+3}}\left(\frac{x}{2}\right)^4 - \frac{1}{3!\overline{|n+4}}\left(\frac{x}{2}\right)^6 + \ldots\right\}, \quad (n > 0)$$

$$= \sum_{r=0}^{\infty} (-1)^r \left(\frac{x}{2}\right)^{n+2r} \frac{1}{r!\overline{|n+r+1}} \tag{5}$$

and corresponding to (4) we have

$$J_{-n}(x) = \sum_{r=0}^{\infty} (-1)^r \left(\frac{x}{2}\right)^{-n+2r} \frac{1}{r!\overline{|-n+r+1}} \tag{6}$$

which is called the Bessel function of the first kind of order $-n$.

Hence, the complete solution of the Bessel's equation (1) may be expressed in the form

$$y = AJ_n(x) + BJ_{-n}(x)$$

Case II: When $n = 0$, the Bessel's equation (1) takes the form

$$x\frac{d^2y}{dx^2} + \frac{dy}{dx} + xy = 0$$

This is called Bessel's equation of order zero. The two roots of the indicial equation are equal i.e., each $= 0$

From equation (3), putting $n = 0$, we have

$$y = x^m \left[1 - \frac{x^2}{(m+2)^2} + \frac{x^4}{(m+2)^2(m+4)^2} - \frac{x^6}{(m+2)^2(m+4)^2(m+6)^2} + \ldots\right]$$

where $a_0 = 1$

If $m = 0$, the first solution is

$$J_0(x) = 1 - \frac{x^2}{2^2} + \frac{1}{2^2 \cdot 4^2}x^4 - \frac{1}{2^2 \cdot 4^2 \cdot 6^2}x^6 + \ldots$$

$$= \sum_{k=0}^{\infty} \frac{(-1)^k}{(k!)^2} \left(\frac{x}{2}\right)^{2k} \quad \text{since } \overline{|k+1} = k!$$

which is the Bessel function of first kind of order zero.

Now

$$\frac{\partial y}{\partial m} = x^m \log^x \left[1 - \frac{x^2}{(m+2)^2} + \frac{x^4}{(m+2)^2(m+4)^2} + \ldots\right]$$

$$+ x^m \left[\frac{x^2}{(m+2)^2}\left(\frac{2}{m+2}\right) - \frac{x^4}{(m+2)^2(m+4)^2}\left(\frac{2}{m+2} + \frac{2}{m+4}\right) + \ldots\right]$$

The second independent solution is given by $\left(\frac{\partial y}{\partial m}\right)_m = 0$

$$= J_0(x) \log^x + \left[\frac{1}{2^2}x^2 - \frac{1}{2^2 \cdot 4^2}\left(1 + \frac{1}{2}\right)x^4 + \frac{1}{2^2 \cdot 4^2 \cdot 6^2}\left(1 + \frac{1}{2} + \frac{1}{3}\right)x^6 - \ldots\right]$$

$$= J_0(x) \log^x + \left[\left(\frac{x}{2}\right)^2 - \frac{1}{(2!)^2}\left(1 + \frac{1}{2}\right)\left(\frac{x}{2}\right)^2 + \frac{1}{(3!)^2}\left(1 + \frac{1}{2} + \frac{1}{3}\right)\left(\frac{x}{2}\right)^6 + \ldots\right]$$

$$= J_0(x) \log^x + \sum_{k=1}^{\infty} \frac{(-1)^{k+1}}{(k!)^2}\left(1 + \frac{1}{2} + \frac{1}{3} + \ldots + \frac{1}{k}\right)\left(\frac{x}{2}\right)^{2k}$$

It is denoted by $Y_0(k)$ and is called Bessel function of the second kind of order zero or *Neumann function*.

Thus, the complete solution of the Bessel's equation of order zero is

$$y = AJ_0(x) + BY_0(x)$$

Case III: When n is an integral Y_2 fails to give a solution for positive values of n and Y_1 fails to give a solution for negative values. Let us find an independent solution of Bessel's equation (1) when n is an integer.

Let $y = u(x)J_n(x)$ be a solution of (1) when n is integral.

Then $\frac{dy}{dx} = u'J_n + uJ_n'$ and $\frac{d^2y}{dx^2} = u''J_n + 2u'J_n' + uJ_n''$.

Substituting the values of y, $\frac{dy}{dx}$, $\frac{d^2y}{dx^2}$ in (1), we get

$$x^2\left(u''J_n + 2u'J_n' + uJ_n''\right) + x\left(u'J_n + uJ_n'\right) + (x^2 - n^2)uJ_n = 0$$

$$\therefore \quad u\left[x^2J_n'' + xJ_n' + (x^2 - n^2)J_n\right] + x^2u''J_n + 2x^2u'J_n' + xu'J_n = 0$$

$$\therefore \quad x^2u''J_n + 2x^2u'J_n' + xu'J_n = 0 \quad \text{since } J_n \text{ is a solution of (1)}$$

Dividing throughout by $x^2u'J_n$ we get

$$\frac{u''}{u'} + 2\frac{J_n'}{J_n} + \frac{1}{x} = 0$$

$$\text{or} \quad \frac{d}{dx}\left[\log u'\right] + 2\frac{d}{dx}\left[\log J_n\right] + \frac{d}{dx}\left[\log x\right] = 0$$

$$\text{or} \quad \frac{d}{dx}\left[\log u' + 2\log J_n + \log x\right] = 0$$

$$\text{or} \quad \frac{d}{dx}\left[\log\left(u'J_n^2x\right)\right] = 0$$

Integrating w.r.t. x, we get

$$\log(u'J_n^2x) = \log B$$

$$\text{or} \quad u'J_n^2x = B$$

$$\therefore \quad u' = \frac{B}{xJ_n^2} \quad \text{or} \quad u = B\int\frac{dx}{xJ_n^2} + A$$

Substituting the value of u in $y = u(x)J_n(x)$, we have

$$y = \left[B\int\frac{dx}{xJ_n^2(x)} + A\right]J_n(x)$$

$$\therefore \quad y = AJ_n(x) + BY_n(x)$$

where $Y_n(x) = J_n(x) = \int\frac{dx}{xJ_n^2(x)}$.

The function $Y_n(x)$ is called the Bessel function of the second kind of order n or Neumann function.

(1) Prove that $J_{-n}(x) = (-1)^nJ_n(x)$.

Proof. We have

$$J_{-n}(x) = \sum_{r=0}^{\infty}(-1)^r\left(\frac{x}{2}\right)^{-n+2r}\frac{1}{r!\,|\overline{-n+r+1}}$$

Put $k = -n + r$ i.e., $r = k + n$ and we know that $\Gamma(k + 1) = k!$ we have

$$J_{-n}(x) = \sum_{k=0}^{\infty} (-1)^{k+n} \frac{\left(\frac{x}{2}\right)^{2k+n}}{(k+n)!k!}$$

$$= (-1)^n \sum_{k=0}^{\infty} \frac{(-1)^k \left(\frac{x}{2}\right)^{n+2k}}{k!\,\Gamma(k+n+1)}$$

$$= (-1)^n J_n(x)$$

$$\therefore \quad J_{-n}(x) = (-1)^n J_n(x)$$

(2) Expansion for J_0 and J_1

Consider

$$J_n(x) = \left(\frac{x}{2}\right)^n \left[\frac{1}{\Gamma(n+1)} - \frac{1}{1!\,\Gamma(n+2)}\left(\frac{x}{2}\right)^2 + \frac{1}{2!\,\Gamma(n+3)}\left(\frac{x}{2}\right)^4 + \ldots\right], \quad (n > 0)$$

Put $n = 0$

$$J_0(x) = 1 - \frac{1}{1!}\left(\frac{x}{2}\right)^2 + \frac{1}{(2!)^2}\left(\frac{x}{2}\right)^4 - \frac{1}{(3!)^2}\left(\frac{x}{2}\right)^6 + \ldots \quad (1)$$

And put $n = 1$

$$J_1(x) = \frac{x}{2}\left[1 - \frac{1}{1!\,2!}\left(\frac{x}{2}\right)^2 + \frac{1}{2!\,3!}\left(\frac{x}{2}\right)^4 - \frac{1}{3!\,4!}\left(\frac{x}{2}\right)^6 + \ldots\right] \quad (2)$$

(3) Prove that

(i) $J_{\frac{-1}{2}}(x) = \sqrt{\frac{2}{\pi x}} \cos x$ (ii) $J_{\frac{1}{2}}(x) = \sqrt{\frac{2}{\pi x}} \sin x$.

Proof. Consider

$$J_n(x) = \left(\frac{x}{2}\right)^n \left[\frac{1}{\Gamma(n+1)} - \frac{1}{1!\,\Gamma(n+2)}\left(\frac{x}{2}\right)^2 + \frac{1}{2!\,\Gamma(n+3)}\left(\frac{x}{2}\right)^4 - \ldots\right] \quad (1)$$

Put $n = \frac{-1}{2}$ in (1)

$$J_{\frac{-1}{2}}(x) = \left(\frac{x}{2}\right)^{\frac{-1}{2}} \left[\frac{1}{\Gamma\frac{1}{2}} - \frac{1}{1!\,\Gamma\frac{3}{2}}\left(\frac{x}{2}\right)^2 + \frac{1}{2!\,\Gamma\frac{5}{2}}\left(\frac{x}{2}\right)^4 - \ldots \right]$$

$$= \frac{x^{\frac{-1}{2}}}{2^{\frac{-1}{2}}} \left[\frac{1}{\sqrt{\frac{1}{2}}} - \frac{1}{1!\frac{1}{2}\Gamma\frac{1}{2}}\left(\frac{x}{2}\right)^2 + \frac{1}{2!\frac{3}{2}\cdot\frac{1}{2}\Gamma\frac{1}{2}}\left(\frac{x}{2}\right)^4 - \ldots \right]$$

$$= \frac{\sqrt{2}}{\sqrt{x}\,\Gamma\frac{1}{2}} \left[1 - \frac{x^2}{1\cdot 2} + \frac{x^4}{1\cdot 2\cdot 3\cdot 4} - \frac{x^6}{1\cdot 2\cdot 3\cdot 4\cdot 5\cdot 6} + \ldots \right]$$

$$= \sqrt{\frac{2}{\pi x}} \left[1 - \frac{x^2}{2!} + \frac{x^4}{4!} - \frac{x^6}{6!} + \ldots \right] \quad \left(\because \Gamma\frac{1}{2} = \sqrt{\pi}\right)$$

$$= \sqrt{\frac{2}{\pi x}} \cos x$$

Again, put $n = \frac{1}{2}$ in (1)

$$J_{\frac{1}{2}}(x) = \left(\frac{x}{2}\right)^{\frac{1}{2}} \left[\frac{1}{\Gamma\frac{3}{2}} - \frac{1}{1!\,\Gamma\frac{5}{2}}\left(\frac{x}{2}\right)^2 + \frac{1}{2!\,\Gamma\frac{7}{2}}\left(\frac{x}{2}\right)^4 - \ldots \right]$$

$$= \left(\frac{x}{2}\right)^{\frac{1}{2}} \left[\frac{1}{\frac{1}{2}\Gamma\frac{1}{2}} - \frac{1}{1!\frac{3}{2}\cdot\frac{1}{2}\Gamma\frac{1}{2}}\left(\frac{x}{2}\right)^2 + \frac{1}{2!\frac{5}{2}\cdot\frac{3}{2}\cdot\frac{1}{2}\Gamma\frac{1}{2}}\left(\frac{x}{2}\right)^4 - \ldots \right]$$

$$= \frac{\sqrt{x}}{\sqrt{2}\,\Gamma\frac{1}{2}} \left[\frac{2}{1!} - \frac{2x^2}{3!} + \frac{2x^4}{5!} - \ldots \right]$$

Multiplying the series by $\frac{x}{2}$ and outside by $\frac{2}{x}$ we get

$$= \frac{\sqrt{2}}{\sqrt{x}\sqrt{\pi}}\left[\frac{x}{1!} - \frac{x^3}{3!} + \frac{x^5}{5!} - \ldots\right] \quad \left(\because \ \left\lceil\frac{1}{2} = \sqrt{\pi}\right)$$

$$= \sqrt{\frac{2}{\pi x}}\sin x$$

Note: Squaring and adding we get

$$J^2_{\frac{1}{2}} + J^2_{\frac{-1}{2}} = \frac{2}{\pi x} \qquad \left(\because \quad \sin^2 x + \cos^2 x = 1\right)$$

5.2.1 Equations Reducible to Bessel's Equation

(1) Reduce the differential equation

$$x^2y'' + xy' + (\lambda^2x^2 - n^2)y = 0 \tag{1}$$

to Bessel's equation. This equation can be reduced to Bessel's equation by putting $z = \lambda x$

i.e., $\frac{dy}{dx} = \frac{dy}{dz}\cdot\frac{dz}{dx} = \lambda\frac{dy}{dz}$ and $\frac{d^2y}{dx^2} = \lambda^2\frac{d^2y}{dz^2}$

$\therefore$ Substituting in the given equation we have

$$z^2\frac{d^2y}{dz^2} + z\frac{dy}{dz} + (z^2 - n^2)y = 0$$

which is the Bessel's equation or order n and its general solution is

$$y = AJ_n(z) + BY_n(z)$$
$$= AJ_n(\lambda x) + BY_n(\lambda x)$$

$\therefore$ $J_n(\lambda x)$ and $Y_n(\lambda x)$ are the solution of (1) which is the modified Bessel's equation.

(2) Reduce the differential equation

$$x\frac{d^2y}{dx^2} + a\frac{dy}{dx} + k^2xy = 0 \tag{2}$$

to Bessel's form.

Put $y = x^n z$ so that $\frac{dy}{dx} = x^n \frac{dz}{dx} + nx^{n-1}z$

and $\frac{d^2y}{dx^2} = x^n \frac{d^2z}{dx^2} + 2n\, x^{n-1} \frac{dz}{dx} + n(n-1)x^{n-2}z$

Substituting in (2) we get

$$x^{n+1}\frac{d^2z}{dx^2} + (2n+a)x^n \frac{dz}{dx} + \left[k^2x^2 + n^2 + (a-1)n\right] x^{n-1}z = 0$$

Dividing by x^{n-1} and put $2n + a = 1$ $\quad\therefore\quad a - 1 = -2n$

$$x^2\frac{d^2z}{dx^2} + x\frac{dz}{dx} + \left(k^2x^2 - n^2\right) z = 0$$

which is the Bessel's equation or order n.

$\therefore$ its solution is

$$y = x^n \left[c_1 J_n(kx) + c_2 J_{-n}(kx)\right]$$

$$= x^n \left[c_1 J_n(kx) + c_2 Y_n(kx)\right]$$

(3) Reduce the differential equation

$$x\frac{d^2y}{dx^2} + c\frac{dy}{dx} + k^2x^r y = 0 \tag{3}$$

to Bessel's form.

Put $x = t^m$ or $t = x^{\frac{1}{m}}$ and equation (3) takes the form

$$t\frac{d^2y}{dt^2} + a\frac{dy}{dt} + (km)^2 y = 0$$

which is similar to (2) and the solution follows as in (2)

Example: Solve the differential equations

(i) $y'' + \frac{y'}{x} + \left(8 - \frac{1}{x^2}\right)y = 0$

(ii) $4y'' + 9xy = 0$

(iii) $y'' + \frac{y'}{x} + \left(1 - \frac{1}{9x^2}\right)y = 0$

(iv) $xy'' + y' + \frac{1}{4}y = 0$

(v) $xy'' + 2y' + \frac{1}{2}xy = 0$

5.2.2 Recurrence Formula for $J_n(x)$

(1) Prove that $\frac{d}{dx}[x^n J_n(x)] = x^n J_{n-1}(x)$

Proof. Consider

$$J_n(x) = \sum_{r=0}^{\infty} (-1)^r \left(\frac{x}{2}\right)^{n+2r} \frac{1}{r!\,\overline{|n+r+1}} \tag{1}$$

Multiplying (1) by x^n

$$x^n J_n(x) = \sum_{r=0}^{\infty} (-1)^r \frac{x^{2(n+r)}}{2^{n+2r} r!\,\overline{|n+r+1}}$$

$$\therefore \quad \frac{d}{dx}[x^n J_n(x)] = \sum_{r=0}^{\infty} (-1)^r \frac{2(n+r)x^{2(n+r)-1}}{2^{n+2r} r!\,\overline{|n+r+1}}$$

$$= x^n \sum_{r=0}^{\infty} (-1)^r \frac{\left(\frac{x}{2}\right)^{n-1+2r}}{r!\,\overline{|n-1+r+1}}$$

$$\left(\because \left(\overline{|n+r+1} = (n+r)\,\overline{|n-1+r-1}\right)\right)$$

$$= x^n J_{n-1}(x)$$

$$\therefore \quad \frac{d}{dx}[x^n J_n(x)] = x^n J_{n-1}(x)$$

(2) Prove that $\frac{d}{dx}[x^{-n} J_n(x)] = -x^{-n} J_{n+1}(x)$.

Proof. We have

$$J_n(x) = \sum_{r=0}^{\infty} (-1)^r \left(\frac{x}{2}\right)^{n+2r} \frac{1}{r!\,\overline{|n+r+1}} \tag{1}$$

Multiplying (1) by x^{-n}

$$\therefore \quad x^{-n} J_n(x) = \sum_{r=0}^{\infty} (-1)^r \frac{x^{2r}}{r!\, 2^{n+2r} \overline{|n+r+1}}$$

Differentiate both sides w.r.t. x

$$\therefore \quad \frac{d}{dx}[x^{-n} J_n(x)] = \sum_{r=0}^{\infty} (-1)^r \frac{2r\, x^{2r-1}}{2^{n+2r} r!\, \overline{|n+r+1}}$$

$$= -x^{-n} \sum_{r=0}^{\infty} (-1)^{r-1} \frac{x^{n+1+2(r-1)}}{2^{n+1+2(r-1)} (r-1)!\, \overline{|n+r+1}}$$

$$= -x^{-n} \sum_{k=0}^{\infty} (-1)^k \frac{\left(\frac{x}{2}\right)^{n+1+2k}}{k!\, \overline{|n+1+k+1}}$$

$$= -x^{-n} J_{n+1}(x)$$

$$\therefore \quad \frac{d}{dx}[x^{-n} J_n(x)] = -x^{-n} J_{n+1}(x)$$

(3) Prove that $J_n(x) = \frac{x}{2n}[J_{n-1}(x) + J_{n+1}(x)]$

Proof. We have

$$\frac{d}{dx}[x^n J_n(x)] = x^n J_{n-1}(x)$$

$$\therefore \quad x^n J_n'(x) + n x^{n-1} J_n(x) = x^n J_{n-1}(x)$$

Dividing by x^n

$$J_n'(x) + \frac{n}{x} J_n(x) = J_{n-1}(x) \qquad \text{(a)}$$

And $\frac{d}{dx}[x^{-n} J_n(x)] = -x^{-n} J_{n+1}(x)$

$$\therefore \quad x^{-n} J_n'(x) - n x^{-n-1} J_n(x) = -x^{-n} J_{n+1}(x)$$

$$\text{or} \quad -J_n'(x) + \frac{n}{x} J_n(x) = J_{n+1}(x) \qquad \text{(b)}$$

Adding (a) and (b) we get

$$\frac{2n}{x}J_n(x) = J_{n-1}(x) + J_{n+1}(x)$$

$$\therefore \quad J_n(x) = \frac{x}{2n}[J_{n-1}(x) + J_{n+1}(x)]$$

(4) Prove that $J_n^1(x) = \frac{1}{2}[J_{n-1}(x) - J_{n+1}(x)]$

Proof. From (3)

$$J_n^1(x) + \frac{n}{x}J_n(x) = J_{n-1}(x) \qquad \text{(a)}$$

$$\text{and} \quad -J_n^1(x) + \frac{n}{x}J_n(x) = J_{n+1}(x) \qquad \text{(b)}$$

Subtracting (b) from (a)

$$2J_n^1(x) = J_{n-1}(x) - J_{n+1}(x)$$

$$\therefore \quad J_n^1(x) = \frac{1}{2}[J_{n-1}(x) - J_{n+1}(x)]$$

(5) Prove that $J_n^1(x) = \frac{n}{x}J_n(x) - J_{n+1}(x)$

Proof. From (3)(b)

$$-J_n^1(x) + \frac{n}{x}J_n(x) = J_{n+1}(x)$$

$$\therefore \quad J_n^1(x) = \frac{n}{x}J_n(x) - J_{n+1}(x)$$

(6) Prove that $J_{n+1}(x) = \frac{2n}{x}J_n(x) - J_{n-1}(x)$

Proof. Again from (3)

$$J_n(x) = \frac{x}{2n}[J_{n-1}(x) + J_{n-1}(x)]$$

$$\therefore \quad \frac{2n}{x}J_n(x) = J_{n-1}(x) + J_{n+1}(x)$$

$$\therefore \quad J_{n+1}(x) = \frac{2n}{x} J_n(x) + J_{n-1}(x)$$

Examples

Example 1: Show that

$$J_{5/2}(x) = \sqrt{\frac{2}{\pi x}} \left(\frac{3 - x^2}{x^2} \sin x - \frac{3}{x} \cos x \right)$$

Solution: We have from recurrence relation

$$J_{n+1}(x) = \frac{2n}{x} J_n(x) - J_{n-1}(x) \tag{1}$$

Put $n = \frac{1}{2}$ in (1)

$$J_{3/2}(x) = \frac{1}{x} J_{1/2}(x) - J_{-1/2}(x)$$

$$= \sqrt{\frac{2}{\pi x}} \left(\frac{\sin x}{x} - \cos x \right)$$

Put $n = \frac{3}{2}$ in (1)

$$\therefore \quad J_{5/2}(x) = \frac{3}{2} J_{3/2}(x) - J_{1/2}(x)$$

$$= \frac{3}{x} \left[\sqrt{\frac{2}{\pi x}} \left(\frac{\sin x}{x} - \cos x \right) \right] - \sqrt{\frac{2}{\pi x}} \sin x$$

$$= \sqrt{\frac{2}{\pi x}} \left[\left(\frac{3 - x^2}{x^2} \right) \sin x - \frac{3}{x} \cos x \right]$$

Example 2: Express $J_4(x)$ in terms of J_0 and J_1

Solution: We have from the recurrence relation

$$J_{n+1}(x) = \frac{2n}{x} J_n(x) - J_{n-1}(x) \tag{1}$$

Put $n = 3$ in (1) we get

$$J_4(x) = \frac{6}{x} J_3(x) - J_2(x)$$

ıt $n = 2$ in (1)

$$J_3(x) = \frac{4}{x}J_2(x) - J_1(x)$$

$$\therefore \quad J_4(x) = \left(\frac{24}{x^2} - 1\right)J_2(x) - \frac{6}{x}J_1(x)$$

Again put $n = 1$ in (1)

$$J_2(x) = \frac{2}{x}J_1(x) - J_0(x)$$

$$\therefore \quad J_4(x) = \left(\frac{24}{x^2} - 1\right)\left[\frac{2}{x}J_1(x) - J_0(x)\right] - \frac{6}{x}J_1(x)$$

$$= \left(\frac{48}{x^3} - \frac{8}{x}\right)J_1(x) - \left(\frac{24}{x^2} - 1\right)J_0(x)$$

Example 3: Express $J_5(x)$ in terms of $J_0(x)$ and $J_1(x)$.

Solution: We have from the recurrence relation

$$J_{n+1}(x) = \frac{2n}{x}J_n(x) - J_{n-1}(x).$$

Putting $n = 1, 2, 3, 4$ successively

$$J_2(x) = \frac{2}{x}J_1(x) - J_0(x)$$

$$J_3(x) = \frac{4}{x}J_2(x) - J_1(x)$$

$$J_4(x) = \frac{6}{x}J_3(x) - J_2(x)$$

$$\text{and} \quad J_5(x) = \frac{8}{x}J_4(x) - J_3(x)$$

Substituting $J_2(x)$ in $J_3(x)$

$$J_3(x) = \frac{4}{x}\left[\frac{2}{x}J_1(x) - J_0(x)\right] - J_1(x)$$

$$= \left(\frac{8}{x^2} - 1\right)J_1(x) - \frac{4}{x}J_0(x)$$

Substituting $J_3(x)$ and $J_2(x)$ in $J_4(x)$ we get

$$J_4(x) = \left(\frac{48}{x^3} - \frac{8}{x}\right) J_1(x) + \left(1 - \frac{24}{x^2}\right) J_0(x).$$

Finally substituting $J_4(x)$ and $J_3(x)$ in $J_5(x)$ we get

$$J_5(x) = \left(\frac{384}{x^4} - \frac{72}{x^2} - 1\right) J_1(x) + \left(\frac{12}{x} - \frac{192}{x^2}\right) J_0(x).$$

Example 4: Prove that $J_n''(x) = \frac{1}{4}[J_{n-2}(x) - 2J_n(x) + J_{n+2}(x)]$

Proof. We have

$$J_n'(x) = \frac{1}{2}[J_{n-1}(x) - J_{n+1}(x)] \tag{1}$$

Differentiate (1)

$$J_n''(x) = \frac{1}{2}\left[J_{n-1}^1(x) - J_{n+1}^1(x)\right] \tag{2}$$

Changing n to $(n-1)$ in (1)

$$J_{n-1}'(x) = \frac{1}{2}[J_{n-2}(x) - J_n(x)] \tag{3}$$

Changing n to $(n+1)$ in (1)

$$J_{n+1}'(x) = \frac{1}{2}[J_n(x) - J_{n+2}(x)] \tag{4}$$

Substituting the values of $J_{n-1}'(x)$ and $J_{n+1}'(x)$ in (2) we get

$$J_n''(x) = \frac{1}{4}[J_{n-2}(x) - 2J_n(x) + J_{n+2}(x)] \tag{5}$$

5.2.3 Orthogonality of Bessel Function

Prove that

$$\int_0^1 xJ_n(x)J_n(\beta x)dx = \begin{cases} 0 & \alpha \neq \beta \\ \frac{1}{2}[J_{n+1}(x)]^2, & \alpha = \beta \end{cases}$$

where α and β are the roofs of $J_n(x) = 0$.

Proof. We know that the solutions of the equations

$$x^2u'' + xu' + \left(\alpha^2x^2 - n^2\right)u = 0 \quad (1)$$

$$\text{and} \quad x^2v'' + xv' + \left(\beta^2x^2 - n^2\right)v = 0 \quad (2)$$

are $u = J_n(\alpha x)$ and $v = J_n(\beta x)$ respectively.

Multiply (1) by $\frac{v}{x}$ and (2) by $\frac{u}{x}$ and then subtracting we get

$$x(u''v - uv'') + (u'v - uv') + (\alpha^2 - \beta^2)xuv = 0$$

$$\text{or} \quad \frac{d}{dx}[x(u'v - uv')] = (\beta^2 - \alpha^2)xuv$$

Now, integrating both sides from 0 to 1

$$\text{i.e.,} \quad \left(\beta^2 - \alpha^2\right)\int_0^1 xuvdx = \left[x(u^1v - uv^1)\right]_0^1$$

$$= \left[u^1v - uv^1\right]_{x=1} \quad (3)$$

Since $u = J_n(\alpha x)$ and $v = J_n(\beta x)$ we have

$$u^1 = J_n^1(\alpha x) \cdot \alpha \quad \text{and} \quad v^1 = J_n^1(\beta x) \cdot \beta$$

Substituting these values in (3) we get

$$\int_0^1 xJ_n(\alpha x)J_n(\beta x)dx = \frac{\alpha J_n^1(\alpha)J_n(\beta) - \beta J_n(\alpha)J_n^1(\beta)}{\beta^2 - \alpha^2} \quad (4)$$

If α and β are distinct roots of $J_n(x) = 0$ then $J_n(\alpha) = J_n(\beta) = 0$. Then equation (4) reduces to

$$\int_0^1 xJ_n(\alpha x)J_n(\beta x)dx = 0 \quad (5)$$

This is known as the orthogonality relation of Bessel functions.

When $\beta = \alpha$, the RHS of (4) is of $\frac{0}{0}$ form. Its value can be found by considering α as a root of $J_n(x) = 0$ and β as variable approaching α.

Then from (4)

$$\lim_{\beta\to\alpha}\int_0^1 xJ_n(\alpha x)J_n(\beta x)dx$$

$$= \lim_{\beta\to\alpha}\frac{\alpha J_n^1(\alpha)J_n(\beta)-0}{\beta^2-\alpha^2}$$

$$\left(\because \quad J_n(\alpha)=0, \alpha \text{ is a root and } 2^{\text{nd}} \text{ term is zero}\right)$$

or by L'Hopital rule

$$\int_0^1 xJ_n^2(\alpha x)dx = \lim_{\beta\to\alpha}\frac{\alpha J_n^1(\alpha)J_n^1(\beta)}{2\beta-0}$$

$$= \frac{1}{2}\left[J_n^1(\alpha)\right]^2 \qquad (\because \quad \beta\to\alpha)$$

$$= \frac{1}{2}[J_{n+1}(\alpha)]^2.$$

Cor : If the integral is of the form 0 to a, it can be shown that

$$\int_0^a xJ_n^2(\alpha x)dx = \frac{a^2}{2}J_{n+1}^2(a\alpha)$$

Where a's are the roots of $J_n(a\alpha)=0$.

5.2.4 Generating Function for $J_n(x)$

Prove that $e^{\frac{1}{2}x\left(t-\frac{1}{t}\right)} = \sum_{n=-\infty}^{\infty} t^n J_n(x)$.

Proof. Consider

$$e^{\frac{1}{2}x\left(t-\frac{1}{t}\right)} = e^{\frac{xt}{2}}\cdot e^{\frac{-x}{2t}}$$

$$= \left[1+\frac{xt}{2}+\frac{1}{2!}\left(\frac{xt}{2}\right)^2+\frac{1}{3!}\left(\frac{xt}{2}\right)^3+\ldots\right]$$

$$\times\left[1-\frac{x}{2t}+\frac{1}{2!}\left(\frac{x}{2t}\right)^2+\frac{1}{3!}\left(\frac{x}{2t}\right)^3+\ldots\right]$$

he coefficient of t^n in this product is

$$= \frac{1}{n!}\left(\frac{x}{2}\right)^n - \frac{1}{1!(n+1)!}\left(\frac{x}{2}\right)^{n+2} + \frac{1}{2!(n+2)!}\left(\frac{x}{2}\right)^{n+4} \cdots$$

s all the integral powers of t, both positive and negative occur we have

$$e^{\frac{1}{2}x\left(t-\frac{1}{t}\right)} = J_0(x) + tJ_1(x) + t^2J_2(x) + t^3J_3(x) + \ldots$$

$$+ t^{-1}J_{-1}(x) + t^{-2}J_{-2}(x) + t^{-3}J_{-3}(x) + \ldots$$

$$= \sum_{n=-\infty}^{\infty} t^n J_n(x)$$

This shows that Bessel's functions of various orders can be derived as coefficient of different powers of t in the expansion of $e^{\frac{1}{2}x\left(t-\frac{1}{t}\right)}$. For this reason it is known as generating function of Bessel functions.

Example 1: Show that $J_n(x) = \frac{1}{\pi}\int_0^{\pi} \cos(n\theta - x\sin\theta)\,d\theta$ where n is an integer.

Solution: We have

$$e^{\frac{1}{2}x\left(t-\frac{1}{t}\right)} = J_0(x) + tJ_1(x) + t^2J_2(x) + \ldots$$

$$+ t^{-1}J_{-1}(x) + t^{-2}J_{-2}(x) + \ldots$$

Since $J_{-n}(x) = (-1)^n J_n(x)$ we have

$$e^{\frac{1}{2}x\left(t-\frac{1}{t}\right)} = J_0 + J_1\left(t - \frac{1}{t}\right) + J_2\left(t^2 + \frac{1}{t^2}\right) + J_3\left(t^3 - \frac{1}{t^3}\right) + \ldots \quad (1)$$

Put $t = \cos\theta + i\sin\theta$

$$\therefore \quad t^p = \cos p\theta + i\sin p\theta \quad \text{and} \quad t^{-p} = \cos p\theta - i\sin p\theta$$

$$\therefore \quad t^p + \frac{1}{t^p} = 2\cos p\theta \quad \text{and} \quad t^p - \frac{1}{t^p} = 2i\sin p\theta.$$

Substituting in (1) we get

$$e^{ix\sin\theta} = J_0 + 2[J_2\cos 2\theta + J_4\cos 4\theta + \cdots]$$

$$+ 2i[J_1\sin\theta + J_3\sin 3\theta + \cdots] \quad (2)$$

Since $e^{ix\sin\theta} = \cos(x\sin\theta) + i\sin(x\sin\theta)$

Equating real and imaginary parts in (2) we get

$$\cos(x\sin\theta) = J_0 + 2\,[J_2\cos 2\theta + J_4\cos 4\theta + \cdots] \qquad (3)$$

$$\sin(x\sin\theta) = 2\,[J_1\sin\theta + J_3\sin 3\theta + \cdots] \qquad (4)$$

which are known as Jacobi series.

Now multiply (3) by $\cos n\theta$ and (4) by $\sin n\theta$ and integrating each of the resulting expressions between 0 and π, we get

$$\frac{1}{\pi}\int_0^{\pi} \cos(x\sin\theta)\cos n\theta d\theta = \begin{cases} J_n(x) & n \text{ is even or } 0 \\ 0 & n \text{ is odd.} \end{cases}$$

and

$$\frac{1}{\pi}\int_0^{\pi} \sin(x\sin\theta)\sin n\theta d\theta = \begin{cases} 0 & n \text{ is even} \\ J_n(x) & n \text{ is odd.} \end{cases}$$

Hence, generally if n is a $+$ ve integer then

$$J_n(x) = \frac{1}{\pi}\int_0^{\pi} [\cos(x\sin\theta)\cos n\theta + \sin(x\sin\theta)\sin n\theta]\, d\theta$$

$$= \frac{1}{\pi}\int_0^{\pi} \cos(n\theta - x\sin\theta) d\theta.$$

Example 2: Prove that $J_0(x) = \frac{1}{\pi}\int_0^{\pi} \cos(x\cos\phi)d\phi.$

Solution: Consider the Jacobi series

$$\cos(x\sin\theta) = J_0 + 2\,[J_2\cos 2\theta + J_4\cos 4\theta + \cdots] \qquad (1)$$

Changing θ to $\frac{\pi}{2} - \phi$ in (1) we have

$$\cos(x\cos\phi) = J_0 + 2J_2\cos(\pi - 2\phi) + 2J_4\cos(2\pi - 4\phi) + \cdots$$

$$= J_0 - 2J_2\cos 2\phi + 2J_4\cos 4\phi + \cdots$$

]egrating w.r.t. ϕ from 0 to π we get

$$\int_0^{\pi} \cos(x\cos\phi)d\phi = \int_0^{\pi} (J_0(x) - 2J_2\cos 2\phi + 2J_4\cos 4\phi + \cdots)\, d\phi$$

$$= J_0(x)\phi - 2J_2(x)\frac{\sin 2\phi}{2} + 2J_4(x)\frac{\sin 4\phi}{4} + \ldots\Big]_0^{\pi}$$

$$= J_0(x)\pi$$

$$\therefore \quad J_0(x) = \frac{1}{\pi}\int_0^{\pi} \cos(x\cos\phi)d\phi.$$

Example 3: For $a > 0$ show that $\int_0^{\infty} e^{-ax}J_0(bx) = \frac{1}{\sqrt{a^2+b^2}}$.

Hence evaluate $\int_0^{\infty} \cos ax\, J_0(bx)dx$.

Solution: Consider the Bessel's integral

$$J_n(x) = \frac{1}{\pi}\int_0^{\pi} \cos(n\theta - x\sin\theta)d\theta$$

Put $n = 0$ and replace x by bx

$$\therefore \quad J_0(bx) = \frac{1}{\pi}\int_0^{\pi} \cos(bx\sin\theta)d\theta$$

Multiply both sides by e^{-ax} and integrate w.r.t. x between the limits $x = 0$ and $x = \infty$.

$$\int_0^{\infty} e^{-ax}J_0(bx)dx = \frac{1}{\pi}\int_0^{\infty}\int_0^{\pi} e^{-ax}\cos(bx\sin\theta)d\theta\, dx$$

To evaluate the double integral, first integrate w.r.t. x and then w.r.t. θ.

$$= \frac{1}{\pi}\int_0^{\pi}\int_0^{\infty} e^{-ax}\cos(bx\sin\theta)dx\, d\theta$$

$$= \frac{1}{\pi} \int_0^{\pi} \frac{e^{-ax}}{a^2 + b^2 \sin\theta} \left[-a\cos(bx\sin\theta) + b\sin\theta\sin(bx\sin\theta) \right]_0^{\infty} d\theta$$

$$= \frac{1}{\pi} \int_0^{\pi} \left[0 - \frac{(-a)}{a^2 + b^2 \sin^2\theta} \right] d\theta$$

$$= \frac{1}{\pi} \int_0^{\pi} \frac{a}{a^2 + b^2 \sin^2\theta} d\theta$$

$$= \frac{2}{\pi} \int_0^{\pi/2} \frac{(a)}{a^2 + b^2 \sin^2\theta} d\theta$$

Put $\tan\theta = t \quad \therefore \quad d\theta = \dfrac{dt}{1 + t^2}$

$$\therefore \quad \sin^2\theta = \frac{\tan^2\theta}{1 + \tan^2\theta} = \frac{t^2}{1 + t^2}$$

$$= \frac{2}{\pi} \int_0^{\infty} \frac{a}{a^2 + \dfrac{b^2 + t^2}{1 + t^2}} \frac{dt}{1 + t^2}$$

$$= \frac{2}{\pi} \int_0^{\infty} \frac{adt}{(a^2 + b^2)t^2 + a^2}$$

$$= \frac{2a}{\pi(a^2 + b^2)} \int_0^{\infty} \frac{1}{\dfrac{a^2}{a^2 + b^2} + t^2} dt$$

$$= \frac{2a}{\pi(a^2 + b^2)} \frac{\sqrt{a^2 + b^2}}{a} \left[\tan^{-1} \frac{\sqrt{a^2 + b^2}}{a} t \right]_0^{\infty}$$

$$= \frac{2}{\pi\sqrt{(a^2 + b^2)}} (\tan^{-1}\infty - \tan^{-1} 0)$$

$$= \frac{2}{\pi\sqrt{(a^2 + b^2)}} \cdot \frac{\pi}{2}$$

$$\therefore \quad \int_0^{\infty} e^{-ax} J_0(bx) dx = \frac{1}{\sqrt{a^2 + b^2}}.$$

(ii) Replace a by ai

$$\therefore \quad \int_0^\infty e^{-iax} J_0(bx)\, dx = \frac{1}{\sqrt{b^2 - a^2}}$$

$$\therefore \quad \int_0^\infty (\cos ax - i \sin ax) J_0(bx)\, dx = \frac{1}{\sqrt{b^2 - a^2}}$$

Equating real part we get

$$\int_0^\infty \cos ax\, J_0(bx)\, dx = \frac{1}{\sqrt{b^2 - a^2}}$$

5.3 LEGENDRE'S EQUATION

The differential equation of the form

$$(1 - x^2)\frac{d^2y}{dx^2} - 2x\frac{dy}{dx} + n(n+1)y = 0 \qquad (1)$$

is called Legendre's equation. Here n is a real number. But in most of the applications only integral values of n are required.

Any solution of equation (1) is called a Legendre's function.

Let $y = a_0 x^m + a_1 x^{m+1} + a_2 x^{m+2} \ldots (2)$ and $a_0 \neq 0$ be the solution of equation (1)

Differentiate (2) and obtain $\frac{dy}{dx}$ and $\frac{d^2y}{dx^2}$

$$\therefore \quad \frac{dy}{dx} = ma_0 x^{m-1} + (m+1)a_1 x^m + (m+2)a_2 x^{m+1} + \ldots$$

$$\frac{d^2y}{dx^2} = m(m-1)a_0 x^{m-2} + (m+1)m\, a_1 x^{m-1} + (m+2)(m+1)a_2 x^m + \ldots$$

Substituting y, $\frac{dy}{dx}$ and $\frac{d^2y}{dx^2}$ in (1) we get

$$(1 - x^2)\left[a_0 m(m-1)x^{m-2} + a_1(m+1)m\, x^{m-1} + \ldots\right]$$

$$-2x\left[a_0 m x^{m-1} + (m+1)a_1 x^m + \ldots\right]$$

$$+ n(n+1)\left[a_0 x^m + a_1 x^{m+1} + \ldots\right] = 0.$$

$$\therefore \quad a_0 m(m-1)x^{m-2} + a_1(m+1)mx^{m-1} + \ldots + [a_{r+2}(m+r+2)(m+r+1)$$
$$- \{(m+r)(m+r+1) - n(n+1)\}ar]x^{m+r} + \ldots = 0$$

Equating to zero the co-efficient of the lowest power of x, i.e., of x^{m-2} we get

$$a_0\, m(m-1) = 0 \quad \therefore \quad m = 0,\ m = 1 \quad [\because \quad a_0 \neq 0]$$

Equating to zero the coefficients of x^{m-1} and x^{m+r} we get

$$a_1(m+1)m = 0 \tag{3}$$

$$a_{r+2}(m+r+2)(m+r+1) - \{(m+r)(m+r+1) - n(n+1)\}a_r = 0 \tag{4}$$

when $m = 0$, (3) is satisfied and therefore, $a_1 \neq 0$

Then (4) gives, taking $r = 0, 1, 2, 3 \ldots$

$$a_2 = \frac{-n(n+1)}{2!} a_0$$

$$a_3 = \frac{-(n-1)(n+2)}{3!} a_1$$

$$a_4 = \frac{-(n-2)(n+3)}{4.3} a_2$$

$$= \frac{n(n-2)(n+1)(n+3)}{4!} a_0$$

$$a_5 = \frac{-(n-3)(n+4)}{5.4} a_3$$

$$= \frac{(n-1)(n-3)(n+2)(n+4)}{5!} a_1$$

Hence, for $m = 0$, there are two independent solutions of (1)

$$y_1 = a_0\left[1 - \frac{n(n+1)}{2!}x^2 + \frac{(n-2)n(n+1)(n+3)}{4!}x^4 + \ldots\right] \tag{5}$$

$$y_2 = a_1\left[x - \frac{(n-1)(n+2)}{3!}x^3 + \frac{(n-3)(n-1)(n+2)(n+4)}{5!}x^5 + \ldots\right] \tag{6}$$

hen $m = 1$ (3) gives $a_1 = 0$

$$\therefore \quad a_3 = a_5 = a_7 = \ldots = 0 \quad \text{and}$$

$$a_2 = -\frac{(n-1)(n+2)}{3!} a_0$$

$$a_4 = \frac{(n-3)(n-1)(n+2)(n+4)}{5!} x^5 \text{ etc.}$$

Thus, for $m = 1$ we get the solution (6) again.

Hence, $y = y_1 + y_2$ is the general solution of (1).

If n is a positive even integer, the series (5) terminates at the term in x^n and y_1 becomes a polynomial. Similarly, if n is an odd integer (6) becomes a polynomial of degree n. Thus whenever n is a positive integer, the general solution of (1) consists of a polynomial solution and an infinite series solution.

These polynomial solutions with a_0 or a_1 so chosen that the value of the polynomial is 1 for $x = 1$ are called *Legendre's Polynomials of order n* and are denoted by $P_n(x)$. The infinite series solution with (a_0 or a_1 properly chosen) is called *Legendre function of the second kind* and is denoted by $\phi_n(x)$.

Thus $P_n(1) = 1$

The general solution of (1) converges in $-1 < x < 1$

$(\because \quad y_1$ and y_2 are convergent for $|x| < 1)$

5.4 RODRIGUE'S FORMULA

Prove that

$$P_n(x) = \frac{1}{n!\, 2^n} \frac{d^n}{dx^n}(x^2 - 1)^n \tag{1}$$

Proof.

$$\text{Let} \quad V = (x^2 - 1)^n$$

$$\text{Then} \quad \frac{dv}{dx} = v_1 = 2nx(x^2 - 1)^{n-1}$$

$$\text{i.e.,} \quad v_1 = \frac{2nx(x^2 - 1)^n}{(x^2 - 1)}$$

$$\therefore \quad (1 - x^2)v_1 + 2nxv = 0 \tag{2}$$

Differentiating (2), $(n+1)$ times by Leibnitz's theorem

$$(1-x^2)v_{n+2} + (n+1)(-2x)v_{n+1} + \frac{1}{21}(n+1)n(-2)v_n$$
$$+ 2n\left[xv_{n+1} + (n+1)v_n\right] = 0$$

$$\therefore \quad (1-x^2)\frac{d^2}{dx^2}(v_n) - 2x\frac{d}{dx}(v_n) + n(n+1)v_n = 0$$

which is Legendre's equation and cv_n is its solution.

Also its finite series solution is $P_n(x)$

$$\therefore \quad P_n(x) = cv_n = c\frac{d^n}{dx^n}(x^2-1)^n \qquad (3)$$

To determine the constant c, Put $x = 1$ in (3)

Then

$$1 = c\left[\frac{d^n}{dx^n}(x-1)^n(x+1)^n\right]_{x=1} \qquad (\because \quad P_n(1) = 1)$$

$$= c\,[n!\,(x+1)^n + \text{terms containing } (x-1) \text{ and its powers}]_{x=1}$$

$$= cn!\,2^n$$

$$\therefore \quad c = \frac{1}{n!\,2^n}$$

Substituting this value of c in (3) we get

$$P_n(x) = \frac{1}{n!\,2^n}\frac{d^n}{dx^n}(x^2-1)^n$$

which is known as Rodrigue's formula.

5.5 LEGENDRE POLYNOMIALS

We have $P_n(x) = \dfrac{1}{n!\,2^n}\dfrac{d^n}{dx^n}(x^2-1)^n$

Put $n = 0$ $\qquad P_0(x) = 1$

$n = 1$ $\qquad P_1(x) = x$

$n = 2$ $\qquad P_2(x) = \frac{1}{2}(3x^2-1)$

$n=3 \qquad P_3(x)=\frac{1}{2}(5x^3-3x)$

$n=4 \qquad P_4(x)=\frac{1}{8}\left(35x^4-30x^2+3\right)$

$n=5 \qquad P_5(x)=\frac{1}{8}\left(63\,x^5-70x^3+15x\right)$ etc.

In general, we have

$$P_n(x)=\sum_{r=0}^{N}\frac{(-1)^r(2n-2r)!}{2^n r!(n-r)!(n-2r)!}x^{n-2r}$$

where $N=\frac{1}{2}n$ or $\frac{1}{2}(n-1)$ according as n is even or odd.

Example 1: Show that $x^4=\frac{1}{35}[8P_4(x)+20P_2(x)+7P_0(x)]$

Solution: We know that

$$P_4(x)=\frac{1}{8}\left(35x^4-30x^2+3\right)$$

$$P_2(x)=\frac{1}{2}(3x^2-1) \quad \text{and} \quad P_0(x)=1$$

$$\therefore \qquad \frac{1}{35}[8P_4(x)+20P_2(x)+7P_0(x)]$$

$$=\frac{1}{35}\left[\left(35\,x^4-30x^2+3\right)+10(3x^2-1)+7\right]$$

$$=x^4$$

$$\therefore \quad x^4=\frac{1}{35}[8P_4(x)+20P_2(x)+7P_0(x)]$$

Example 2: Express $f(x)=x^4+3x^3-x^2+5x-2$ in terms of Legendre polynomials.

Solution: Let $f(x)=x^4+3x^3-x^2+5x-2$

Since $\quad P_4(x)=\frac{35}{8}x^4-\frac{15}{4}x^2+\frac{3}{8}$

$$\therefore \quad x^4=\frac{8}{35}P_4(x)+\frac{6}{7}x^2-\frac{3}{35}$$

$$\therefore \quad f(x) = \left(\frac{8}{35}P_4(x) + \frac{6}{7}x^2 - \frac{3}{35}\right) + 3x^3 - x^2 + 5x + 2$$

$$= \frac{8}{35}P_4(x) + 3x^3 - \frac{1}{7}x^2 + 5x - \frac{73}{35}$$

Put $x^3 = \frac{2}{5}P_3(x) + \frac{3}{5}x$ and $x^2 = \frac{2}{3}P_2(x) + \frac{1}{3}$

$$= \frac{8}{35}P_4(x) + 3\left[\frac{2}{5}P_3(x) + \frac{3}{5}x\right] - \frac{1}{7}\left[\frac{2}{3}P_2(x) + \frac{1}{3}\right] + 5x - \frac{73}{35}$$

$$= \frac{8}{35}P_4(x) + \frac{6}{5}P_3(x) - \frac{2}{21}P_2(x) + \frac{34}{5}x - \frac{224}{105}$$

Now put $x = P_1(x)$ and $1 = P_0(x)$

$$\therefore \quad f(x) = \frac{8}{35}P_4(x) + \frac{6}{5}P_3(x) - \frac{2}{21}P_2(x) + \frac{34}{5}P_1(x) - \frac{224}{105}P_0(x)$$

Example 3: Express $f(n) = x^3 - 5x^2 + x + 2$ in terms of Legendre's polynomials.

Solution: We have $P_3(x) = \frac{1}{2}(5x^3 - 3x)$

$$\therefore \quad x^3 = \frac{2}{5}P_3(x) + \frac{3}{5}x$$

$$\therefore \quad f(x) = \left[\frac{2}{5}P_3(x) + \frac{3}{5}x\right] - 5x^2 + x + 2$$

$$= \frac{2}{5}P_3(x) - 5x^2 + \frac{8}{3}x + 2$$

$$= \frac{2}{5}P_3(x) - 5\left[\frac{2}{3}P_2(x) + \frac{1}{3}\right] + \frac{8}{5}x + 2 \qquad \left(\because \; P_2(x) = \frac{1}{2}(3x^2 - 1), \; x^2 = \frac{2}{3}P_2(x) + \frac{1}{3}\right)$$

$$= \frac{2}{5}P_3(x) - \frac{10}{3}P_2(x) + \frac{8}{5}x + \frac{1}{3}$$

$$\therefore \quad f(x) = \frac{2}{5}P_3(x) - \frac{10}{3}P_2(x) + \frac{8}{5}P_1(x) + \frac{1}{3}P_0(x) \qquad (\because \; x = P_1(x), \; 1 = P_0(x))$$

Example 4: Expand $f(x) = x^3$ in a series of the form $\sum_{n=0}^{\infty} Cr\, P_r(x)$ i.e., find series of polynomials for x^3.

Solution: Let $f(x) = x^3$ is a polynomial of degree 3

Thus from Legendre's series we have

$$x^3 = \sum_{n=0}^{\infty} C_r\, P_r(x)$$

$$= C_0\, P_0(x) + C_1 P_1(x) + C_2 P_2(x) + C_3 P_3(x)$$

$$\text{where} \quad C_r = \left(r + \frac{1}{2}\right) \int_{-1}^{1} x^3 P_r(x)\, dx$$

Now, $P_0(x) = 1$, $P_1(x) = x$, $P_2(x) = \frac{1}{2}(3x^2 - 1)$ and $P_3(x) = \frac{1}{2}(5x^3 - 3x)$.

$$\therefore \quad C_0 = \frac{1}{2} \int_{-1}^{1} x^3 dx = \frac{1}{2} \left[\frac{x^4}{4}\right]_{-1}^{1} = 0$$

$$C_1 = \frac{3}{2} \int_{-1}^{1} x^4\, dx = \frac{3}{2} \left[\frac{x^5}{5}\right]_{-1}^{1} = \frac{3}{5}$$

$$C_2 = \frac{5}{2} \int_{-1}^{1} \frac{1}{2}(3x^2 - 1)x^3\, dx = \frac{5}{4} \int_{-1}^{1} (3x^5 - x^3)\, dx = 0$$

$$C_3 = \frac{7}{3} \int_{-1}^{1} \frac{1}{2}(5x^3 - 3x)x^3\, dx = \frac{2}{5}$$

$$\therefore \quad x^3 = \frac{3}{5} P_1(x) + \frac{2}{5} P_3(x)$$

Example 5: Express $f(x) = x^3 - 5x^2 + 6x + 1$ in terms of Legendre polynomial of degree n.

Example 6: Express $f(x) = x^5 + 1$ in terms of Legendre polynomial of degree n.

Example 7: Express $f(x) = x^3 - 5x^2 + x + 2$ in terms of Legendre polynomial of degree n.

5.5.1 Generating Function for $P_n(x)$

Prove that $(1 - 2xt + t^2)^{-\frac{1}{2}} = \sum_{n=0}^{\infty} t^n P_n(x)$

Proof. We know that

$$(1-z)^{-\frac{1}{2}} = 1 + \frac{1}{2}z + \frac{\frac{1}{2}\,\frac{3}{2}}{2!}z^2 + \frac{\frac{1}{2}\cdot\frac{3}{2}\cdot\frac{5}{2}}{3!}z^3 + \ldots$$

$$= 1 + \frac{2!}{(1!)^2\, 2^2}z + \frac{4!}{(2!)^2\, 2^4}z^2 + \frac{6!}{(3!)^2\, 2^6}z^3 + \ldots$$

Consider

$$\lfloor 1 - t(2x-t)\rfloor^{-\frac{1}{2}} = 1 + \frac{2!}{(1!)^2\, 2^2}\, t(2x-t) + \frac{4!}{(2!)^2\, 2^4}\, t^2(2x-t)^2 + \ldots$$

$$+ \frac{(2n-2r)!}{\lfloor (n-r)!\rfloor^2\, 2^{2n-2r}} t^{n-r}(2x-t)^{n-r} + \ldots$$

$$+ \frac{(2n)!}{(n!)^2\, 2^{2n}}\, t^n (2x-t)^n \tag{1}$$

The term in t^n from the term containing $t^{n-r}(2x-t)^{n-r}$

$$= \frac{(2n-2r)!}{\lfloor (n-r)!\rfloor^2\, 2^{2n-2r}} t^{n-r}\, C_r^{n-r}(-t)^r (2x)^{n-2r}$$

$$= \frac{(2n-2r)!}{\lfloor (n-r)!\rfloor^2\, 2^{2n-2r}}\, \frac{(n-r)!}{r!\,(n-2r)!}(-1)^r\, t^n\, (2x)^{n-2r}$$

$$= \frac{(-1)^r (2n-2r)!}{2^n r!(n-r)!\,(n-2r)!} x^{n-2r}\, t^n$$

Collecting all the terms in t^n which will occur in the term containing $t^n(2x-t)^n$ and the preceding terms we see that terms in t^n

$$= \sum_{r=0}^{N} \frac{(-1)^r\, (2n-2r)!}{2^n r!(n-r)!\,(n-2r)!} x^{n-2r}\, t^n$$

$$= P_n(x)\, t^n$$

where $N = \frac{1}{2}n$ or $\frac{1}{2}(n-1)$ according as n is even or odd.

Hence (1) may be written as

$$\lfloor 1 - t(2x - t) \rfloor^{-\frac{1}{2}} = \sum_{n=0}^{\infty} P_n(x) t^n \tag{2}$$

This shows that $P_n(x)$ is the coefficient of t^n in the expansion of $(1 - 2xt + t^2)^{-\frac{1}{2}}$. That is why, it is known as the generating function of Legendre's polynomial.

Corollary 1: Prove that $P_n(1) = 1$

Proof. Taking $x = 1$ in (2).

$$(1 - 2t + t^2)^{-\frac{1}{2}} = \sum_{n=0}^{\infty} P_n(1) t^n$$

$$\sum_{n=0}^{\infty} P_n(1) t^n = (1 - t)^{-1}$$

$$= 1 + t + t^2 + t^3 + \ldots + t^n + \ldots$$

Equating the coefficient of t^n we get

$$P_n(1) = 1$$

Corollary 2: Prove that $P_n(-1) = (-1)^n$

Proof. Taking $x = -1$ in (2)

$$\sum_{n=0}^{\infty} P_n(-1) t^n = (1 + t)^{-1}$$

$$= 1 - t + t^2 - t^3 + t^4 \ldots + (-1)^n t^n + \ldots$$

Equating the coefficient of t^n we get

$$P_n(-1) = (-1)^n$$

Corollary 3: Prove that $P_{2m}^{(0)} = (-1)^m \dfrac{1\cdot 3\cdot 5\ldots(2m-1)}{2\cdot 4\cdot 6\ldots 2m}$ and $P_{2m+1}^{(0)} = 0$.

Proof. Putting $x = 0$ in (2)

$$\sum_{n=0}^{\infty} P_n(0)t^n = (1+t^2)^{-\frac{1}{2}}$$

$$= 1 - \frac{1}{2}t^2 + \frac{1.3}{2.4}t^4 + \ldots + \frac{(-1)^r\, 1\cdot 3\cdot 5\ldots(2r-1)}{2\cdot 4\cdot 6\ldots 2r}t^{2r} + \ldots$$

Equating the coefficient of t^{2m} we get

$$P_{2m}^{(0)} = (-1)^m \frac{1\cdot 3\cdot 5\ldots(2m-1)}{2\cdot 4\cdot 6\ldots 2m}$$

Similarly equating the coefficient of t^{2m+1} we get

$$P_{2m+1}^{(0)} = 0$$

5.5.2 Recurrence Formulae for $P_n(x)$

(1) Prove that

$$(n+1)P_{n+1}^{(x)} = (2n+1)x\,P_n(x) - nP_{n-1}^{(x)}$$

Proof. We have

$$\left(1 - 2xt + t^2\right)^{-\frac{1}{2}} = \sum_{n=0}^{\infty} P_n(x)\, t^n \qquad (1)$$

Differentiate w.r.t.'t' partially

$$-\frac{1}{2}(1-2xt+t^2)^{-\frac{3}{2}}(-2x+2t) = \sum_{n=0}^{\infty} nP_n(x)t^{n-1}$$

$$\therefore\quad (x-t)(1-2xt+t^2)^{-\frac{1}{2}} = \left(1-2xt+t^2\right)\sum nP_n(x)t^{n-1}$$

$$\therefore\quad (x-t)\sum P_n(x)t^n = (1-2xt+t^2)\sum nP_n(x)t^{n-1}$$

Equating coefficient of t^n from both sides we get

$$xP_n(x) - P_{n-1}(x) = (n+1)P_n(x) - 2nx\,P_n(x) + (n-1)P_{n-1}^{(x)}$$

$$\therefore\quad (n+1)P_{n+1}^{(x)} = (2n+1)xP_n(x) - nP(x)_{n-1}$$

(2) Prove that $nP_n(x) = xP_n^1(x) - P_{n-1}^1(x)$

Proof. Consider

$$(1-2xt+t^2)^{-\frac{1}{2}} = \sum_{n=0}^{\infty} P_n(x)t^n \quad (1)$$

Differentiate partially w.r.t. x

$$-\frac{1}{2}\left(1-2xt+t^2\right)^{-3/2}(-2t) = \sum P_n^1(x)t^n \quad (2)$$

Differentiate again (1) w.r.t.'t' partially

$$(x-t)(1-2xt+t^2)^{-\frac{3}{2}} = \sum nP_n(x) + ^{n-1} \quad (3)$$

Dividing (3) by (2) we get

$$\frac{x-t}{t} = \frac{\sum nP_n(x)t^{n-1}}{\sum P_n^1(x)t^n}$$

$$\therefore \quad \sum nP_n(x)t^n = (x-t)\sum P_n^1(x)t^n$$

Equating the coefficients of t^n from both sides we get

$$nP_n(x) = xP_n^1(x) - P_{n-1}^1(x).$$

(3) Prove that

$$(2n+1)P_n(x) = P_{n+1}^1(x) - P_{n-1}^1(x)$$

Proof. Consider the first recurrence formula

$$(n+1)P_{n+1}(x) = (2n+1)xP_n(x) - nP_{n-1}(x) \quad (1)$$

Differentiate w.r.t. x

$$(n+1)P_{n+1}^1(x) = (2n+1)P_n(x) + (2n+1)xP_n^1(x) - nP_{n-1}^1(x) \quad (a)$$

Also from R.R. (2)

$$nP_n(x) = xP_n^1(x) - P_{n-1}^1(x) \quad (2)$$

Substituting $xP_n^1(x)$ from (2) in (a).

$$(n+1)P_{n+1}^1(x) = (2n+1)P_n(x) + (2n+1)\left[nP_n(x) + P_{n-1}^1(x)\right]$$

$$- nP_{n-1}^1(x)$$

$$\therefore \quad (2n+1)P_n(x) = P_{n+1}^1(x) - P_{n-1}^1(x)$$

(4) Prove that $P_n^1(x) = xP_{n-1}^1(x) + nP_{n-1}(x)$

Proof. Rewriting (a) as

$$(n+1)P_{n+1}^1(x) = (2n+1)P_n(x) + (2n+1)xP_n^1(x) - n\left[xP_n^1(x) - nP_n(x)\right]$$

(by R.R.2)

$$= (2n+1)P_n(x) + (n+1)xP_n^1(x) + n^2P_n(x)$$

$$= (n+1)xP_n^1(x) + (n^2+2n+1)P_n(x)$$

$$\therefore \quad P_{n+1}^1(x) = xP_n^1(x) + (n+1)P_n(x)$$

Replace n by $n-1$ we get

$$P_n^1(x) = xP_{n-1}^1(x) + nP_{n-1}(x)$$

(5) Prove that $(1-x^2)p_n^1(x) = n\,[P_{n-1}(x) - xP_n(x)]$

Proof. Rewriting R.R.S. (2) and (4) as

$$xP_n^1(x) - P_{n-1}^1(x) = nP_n(x) \tag{1}$$

$$\text{and} \quad P_n^1(x) - xP_{n-1}^1(x) = nP_{n-1}(x) \tag{2}$$

Multiply (1) by x and subtracting from (2) we get

$$(1-x^2)P_n^1(x) = n\,[P_{n-1}(x) - xP_n(x)]$$

5.5.3 Orthogonality of Legendre Polynomials

Prove that

$$\int_{-1}^{1} P_m(x)P_n(x)dx = 0,\ m \neq n$$

$$= \frac{2}{2n+1},\ m = n$$

Proof. We know that the solutions of

$$(1-x^2)u'' - 2xu' + m(m+1)u = 0 \tag{1}$$

and $$(1-x^2)v'' - 2xv' + n(n+1)v = 0 \tag{2}$$

Multiplying (1) by v and (2) by u and subtracting we get

$$(1-x^2)(u''v - uv'') - 2x(u'v - uv') + [m(m+1) - n(n+1)]\,uv = 0$$

or $$\frac{d}{dx}\left[(1-x^2)(u'v - uv')\right] + (m-n)(m+n+1)uv = 0$$

Now integrating from -1 to $+1$ we get

$$(m-n)(m+n+1)\int_{-1}^{1} uv dx = \left[(1-x^2)(uv' - u'v)\right]_{-1}^{1} = 0$$

Hence $$\int_{-1}^{1} P_m(x)P_n(x)dx = 0 \tag{3}$$

(Provided $m \neq n$)

This is known as the orthogonal property of Legendre polynomials.

When $m = n$ we have from Rodrigue formula

$$(n!2^n)^2 \int_{-1}^{1} P_n^2(x)dx = \int_{-1}^{1} D^n(x^2-1)^n D^n(x^2-1)^n dx$$

Integrating by parts

$$= D^n(x^2-1)^n D^{n-1}(x^2-1)^n \Big|_{-1}^{1} - \int_{-1}^{1} D^{n+1}(x^2-1)^n D^{n-1}(x^2-1)^n dx$$

Since $D^{n-1}(x^2-1)^n$ has x^2-1 as a factor, the first term on the right vanishes for $x = \pm 1$.

Thus

$$(n!2^n)^2 \int_{-1}^{1} P_n^2(x)dx = -\int_{-1}^{1} D^{n+1}(x^2-1)^n D^{n-1}(x^2-1)^n dx$$

Integrating by parts $(n-1)$ times

$$= (-1)^n \int_{-1}^{1} D^{2n}(x^2-1)^n (x^2-1)^n dx$$

$$= (-1)^n \int_{-1}^{1} (2n)!(x^2-1)^n dx$$

$$= (2n)!(-1)^n(-1)^n 2\int_{0}^{1} (1-x^2)^n dx$$

$$= 2(2n)!\int_{0}^{1} (1-x^2)^n dx$$

Put $x = \sin\theta$

$$= 2(2n)! \int_{0}^{\pi/2} \cos\theta^{2n+1} d\theta.$$

$$= 2(2n)! \frac{2n(2n-2)\ldots4.2}{(2n+1)(2n-1)\ldots3}$$

$$= 2(2n)! \frac{[2n(2n-2)\ldots4.2]^2}{(2n+1)!}$$

$$= \frac{2(2n)!}{2n+1}\frac{(2^n n!)^2}{(2n)!}$$

$$= \frac{2}{2n+1}(2^n n!)^2$$

$$\therefore \quad (n!2^n)^2 \int_{-1}^{1} P_n^2(x)dx = \frac{2}{2n+1}(2^n n!)^2$$

$$\therefore \quad \int_{-1}^{1} P_n^2(x)dx = \frac{2}{2n+1}$$

5.5.4 Legendre's Series for *f*(*x*) when *f*(*x*) is a polynomial

Let $f(x)$ be a polynomial of degree n defined from $x = -1$ to $x = +1$. Then

$$f(x) = \sum_{r=0}^{n} C_r P_r(x) \tag{1}$$

where $C_r = \left(r + \frac{1}{2}\right) \int_{-1}^{1} f(x)P_r(x)dx.$

The series (1) will be convergent and the series expansion is unique provided $f(x)$ is continuous and has continuous derivatives in $-1 \le x \le 1$.

Example 1: If $f(x) = 0 \quad -1 \le x \le 0$
$\qquad = x \quad 0 < x < 1$

Show that

$$f(x) = \frac{1}{4}P_0(x) + \frac{1}{2}P_1(x) + \frac{5}{16}P_2(x) - y\frac{3}{32}P_4(x) + \cdots$$

Solution: Let $f(x) = \sum_{n=1}^{\infty} C_n P_n(x)$

Then

$$C_n = \left(n + \frac{1}{2}\right) \int_{-1}^{1} f(x)P_n(x)dx$$

$$= \left(n + \frac{1}{2}\right) \left[\int_{-1}^{0} (0)P_n(x)dx + \int_{0}^{1} xP_n(x)dx\right]$$

$$= \left(n + \frac{1}{2}\right) \int_{0}^{1} xP_n(x)dx$$

Now

$$C_0 = \frac{1}{2}\int_0^1 xP_0(x)dx = \frac{1}{2}\int_0^1 xdx = \frac{1}{4}$$

$$C_1 = \frac{3}{2}\int_0^1 xP_1(x)dx = \frac{3}{2}\int_0^1 x^2dx = \frac{1}{2}$$

$$C_2 = \frac{5}{2}\int_0^1 xP_2(x)dx = \frac{5}{2}\int_0^1 x\frac{(3x^2-1)}{2}dx$$

$$= \frac{5}{4}\int_0^1 (3x^3 - x)dx = \frac{5}{16}$$

$$C_3 = \frac{7}{2}\int_0^1 xP_3(x)dx = \frac{7}{2}\int_0^1 x\frac{(5x^3-3x)}{2}dx$$

$$= \frac{7}{4}\int_0^1 (5x^4 - 3x^2)dx = 0$$

$$C_4 = \frac{9}{2}\int_0^1 xP_4(x)dx = \frac{9}{2}\int_0^1 x\frac{(35x^4-30x^2+3)}{8}dx$$

$$= \frac{-3}{32}$$

$$\therefore \quad f(x) = \frac{1}{4}P_0(x) + \frac{1}{2}P_1(x) + \frac{5}{16}P_2(x) - \frac{3}{32}P_4(x) + \cdots$$

QUESTION BANK

1. Obtain the solution of Laplace's equation in cylindrical coordinates.
2. Obtain the solution of Laplace's equation in spherical coordinates.
3. Find the potential in the interior of a sphere of unit radius when the potential on the surface is $f(\theta) = \cos^2\theta$.

4. A fluid (wind) with velocity 50 km/hr flows uniformly over a spherical dome having radius of 5 m. The upstream pressure and temperature are equal to those inside the dome. Estimate the upward force on the dome.

5. Obtain the series solution of Bessel's differential equation in the form

$$y = AJ_n(x) + BJ_{-n}(x)$$

6. Show that (i) $J_{1/2}(x) = \sqrt{\frac{2}{\pi x}} \sin x$ (ii) $J_{-1/2}(x) = \sqrt{\frac{2}{\pi x}} \cos x$.

7. Prove that $\int_0^1 x J_n^2(\alpha x) dx = \frac{1}{2} [J_n'(\alpha)]_0^2$, where α is the root of the equation $J_n(x) = 0$.

8. Obtain the Jacobi's series

$$\cos(x \sin\theta) = J_0 + 2[J_2 \cos 2\theta + J_4 \cos 4\theta + \ldots]$$

$$\sin(x \sin\theta) = 2[J_1 \sin\theta + J_3 \sin 3\theta + \ldots]$$

9. Prove that

$$J_0^2 + 2J_1^2 + 2J_2^2 + 2J_3^2 + \ldots = 1$$

10. Prove that

(i) $2nJ_n(x) = x[J_{n+1}(x) + J_{n-1}(x)]$

(ii) $J_n'(x) = \frac{1}{2}[J_{n+1}(x) + J_{n-1}(x)]$

11. Show that

(i) $J_{-n}(x) = (-1)^n J_n(x) \quad n = 1, 2, 3 \ldots$

(ii) $\frac{d}{dx}[x^n J_n(x)] = x^n J_{n-1}(x)$

12. Derive Bessel's integral formula

$$J_n(x) = \frac{1}{\pi} \int_0^{\pi} (\cos n\theta - x \sin\theta) d\theta$$

where n is a positive integer.

13. Express $J_5(x)$ in terms of $J_0(x)$ and $J_1(x)$.

Ans: $J_5(x) = \left[\frac{384}{x^4} - \frac{72}{x^2} - 1\right] J_1(x) + \left[\frac{12}{x} - \frac{192}{x^3}\right] J_0(x).$

14. Prove that

$$J_{5/2}(x) = \sqrt{\frac{2}{\pi x}}\left\{\frac{3 - x^2}{x^2}\sin x - \frac{3}{x}\cos x\right\}$$

15. Prove that $e^{\frac{1}{2}x\left(t - \frac{1}{t}\right)} = \sum_{n=-\infty}^{\infty} t^n J_n(x)$

16. Prove that

$$\int_0^1 x J_n(\alpha x) J_n(\beta x)\,dx = \begin{cases} 0 & \alpha \neq \beta \\ \frac{1}{2}[J_{n+1}(\alpha)]^2, & \alpha = \beta \end{cases}$$

where α, β are the roots of $J_n(x) = 0$.

17. Obtain the series solution of the Legendre's differential equation

$$(1 - x^2)y'' - 2xy' + n(n + 1)y = 0$$

18. Establish the Rodrigue formula

$$P_n(x) = \frac{1}{2^n n!}\frac{d^n}{dx^n}(x^2 - 1)^n$$

19. Express $x^4 + 3x^3 - x^2 + 5x - 2$ in terms of Legendre polynomials.

20. Express $x^3 + 2x^2 - x - 3$ in terms of Legendre polynomials.

21. Compute P_0, P_1, P_2, P_3, P_4 using Rodrigue's formula.

22. Show that $\int_{-1}^{1} P_n^2(x)\,dx = \frac{2}{2n + 1}$.

23. Show that $(1 - 2xz + z^2)^{\frac{-1}{2}} = \sum_{n=0}^{\infty} P_n(x) z^n$.

24. Establish the orthogonal property

$$\int_{-1}^{1} P_m(x) \cdot P_n(x)\,dx = 0, \quad m \neq n$$

where m and n are positive integers.

25. Using generating function for $P_n(x)$ derive the recurrence formulae.

(a) $(n+1)P_{n+1}(x) = (2n+1)xP_n(x) - nP_{n-1}(x)$

(b) $nP_n(x) = xP_n'(x) - P_{n-1}'(x)$

(c) $(2n+1)P_n(x) = P_{n+1}'(x) - P_{n-1}'(x)$

(d) $P_n'(x) = xP_{n-1}'(x) + nP_{n-1}(x)$

(e) $(1-x^2)P_n'(x) = n[P_{n-1}(x) - xP_n(x)]$

26. If $\begin{aligned} f(x) &= 0 \quad -1 < x \le 0 \\ &= x \quad 0 < x < 1 \end{aligned}$

Show that

$$f(x) = \frac{1}{4}P_0(x) + \frac{1}{2}P_1(x) + \frac{5}{16}P_2(x) - \frac{3}{32}P_4(x) + \ldots$$

27. Prove that

$$P_{2m}(0) = (-1)^m \frac{1\cdot 3\cdot 5\ldots(2m-1)}{2\cdot 4\cdot 6\ldots 2m}$$

$$\text{and} \quad P_{2m}+1(0) = 0$$

28. Find the solutions of the following differential equations in terms of Bessel's functions.

(i) $xy'' + y = 0$

Ans: $y = \sqrt{x}\,[C_1J_1(2\sqrt{x}) + C_2Y_1(2\sqrt{x})]$

(ii) $xy'' - y' + 4x^3y = 0$

Ans: $y = x\left[C_1J_1J_{1/2}(x^2) + C_2J_{-1/2}(x^2)\right]$

(iii) $y'' + \dfrac{1}{x}y' + 4\left(1 - \dfrac{1}{x^2}\right)y = 0$

Ans: $y = C_1J_2(2x) + C_2Y_2(2x)$

(iv) $y'' + \dfrac{1}{x}y' + \left(3 - \dfrac{1}{4x^2}\right)y = 0$

Ans: $y = C_1J_{1/2}(\sqrt{3}x) + C_2J_{-1/2}(\sqrt{3}x)$

(v) Show that $x^nJ_n(x)$ is a solution of the equation

$$xy'' + (1-2n)y' + xy = 0$$

UNIT 6

Probability Theory – I

6.1 INTRODUCTION

In day-to-day conversation, we very commonly use the word 'probability' or 'chance' and we have a vague idea about its meaning. For example, we come across statements like "Probably it may rain today", "The chances of teams A and B winning a cricket match are equal", "It is possible that I may not be able to join you at the marriage party". All these statements convey the same sense. i.e., the event is not certain to take place, or in other words, there is uncertainty about the happening of the event in question. Thus, the probability has become a part of our everyday life. In personal and management decisions, we face uncertainty and use probability theory, whether or not we admit the use of something so sophisticated. To quote Levin: "We live in a world in which we are unable to forecast the future with complete certainty. Our need to cope with uncertainty leads us to the study and use of probability theory". So the study of probability provides a mathematical framework for such assertions and is essential in every decision-making process. Before defining probability, let us explain a few terms.

Events: Any finite subset of a sample space is called an event.

Two trivial subsets of S are ϕ and S itself, ϕ and S are also events.

Exhaustive Events: A set of events is said to be exhaustive, if it includes all the possible events.

For example, in tossing a coin there are two exhaustive cases either a head or a tail and there is no third possibility.

Mutually Exclusive: If the occurrence of one of the events precludes the occurrence of all others then such a set of events is said to be mutually exclusive.

For example, as in tossing a coin either head comes up or the tail and both cannot happen at the same time, i.e., they are mutually exclusive cases.

Equally likely: If one of the events cannot be expected to happen in preference to another then such events are said to be equally likely.

For instance, in tossing a coin, the coming up of the head or the tail is equally likely.

Experiment: Any operation resulting in two or more outcomes is known as an experiment.

For example, Tossing of a coin, rolling a die, choosing a defective item from a box etc., are the examples of an experiment.

Sample space: A set representing all possible outcomes of a random experiment is called a sample space and is usually denoted by *S*.

For example, when a pair of coins is tossed the sample space is $\{HH, TH, HT, TT\}$

Thus, the sample space consists of four sample points.

Disjoint sets: Two sets *A* and *B* are said to be disjoint sets if $A \cap B = \phi$.

Models: Models are mathematical descriptions of probability. They involve additive, multiplicative properties to the probability.

Definition: If there are *n* exhaustive, mutually exclusive and equally likely cases of which *m* are favourable to an event *A*, then the probability (P) of happening of *A* is $P(A) = \frac{m}{n}$. As there are $n - m$ cases in which *A* will not happen therefore the chance of *A* not happening is *q* or $P(\bar{A})$ so that

$$q = \frac{n-m}{n} = 1 - \frac{m}{n} = 1 - p$$

$$\therefore \quad P(\bar{A}) = 1 - P(A)$$

$$\therefore \quad P(A) + P(\bar{A}) = 1 \quad \text{or} \quad \boxed{p + q = 1}$$

6.1.1 Axioms of Probability

Let *S* be a sample space of a random experiment and $A \subset S$ then the probability $P(A)$ of an event *A* is a set function satisfying the following axioms.

(1) Axiom of positiveness $P(A) \geq 0$

i.e., the probability of an event ranges from 0 to 1. If the event cannot take place its probability shall be 0 and if it is certain, its probability shall be 1.

(2) Axiom of certainty: $P(S) = 1$.
i.e., the probability of the entire sample space is 1.

(3) Axiom of unity:

$$P\left(\bigcup_{j=1}^{n} A_j\right) = \sum_{j=1}^{n} P(A_j)$$

i.e., $P(A_1 \cup A_2 \cup A_3 \cdots \cup A_n) = P(A_1) + P(A_2) + \ldots + P(A_n)$ where A_1, $A_2, A_3 \ldots A_n$ are disjoint subsets of the sample space S.

6.1.2 Independent and Dependent Events

Two or more events are said to be independent when the outcome of one does not affect, and is not affected by the other.

For example, if a coin is tossed twice, the result of the second throw would in no way be affected by the result of the first throw.

Dependent events are those in which the occurrence or non-occurrence of one event in any one trial affects the probability of other events in other trials.

For example, If a card is drawn from a pack of playing cards and is not replaced, this will alter the probability that the second drawn is, say an ace, i.e., probability of drawing any card is $\frac{1}{52}$ and then the probability of drawing an ace is $= \frac{4}{51}$ as the card drawn first is not replaced. Therefore the probability is affected.

6.1.3 Addition Theorem of Probability

Statement: The probability of occurrence of at least one of the two events A and B is

$$P(A \cup B) = P(A) + P(B) - P(A \cap B)$$

Proof. This theorem, we prove by Venn diagram.

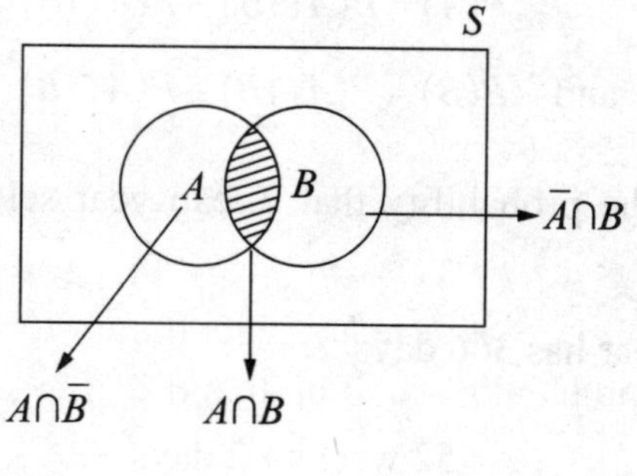

We have

$$A \cup B = A \cup (B \cap \bar{A})$$

$$\text{and } B = (A \cap B) \cup (B \cap \bar{A})$$

$$\therefore \quad P(A \cup B) = P(A) + P(B \cap \bar{A}) \qquad (1)$$

$$\text{and } \quad P(B) = P(A \cap B) + P(B \cap \bar{A}) \qquad (2)$$

Subtracting (2) from (1) we get

$$P(A \cup B) - P(B) = P(A) - P(A \cap B)$$

$$\therefore \quad P(A \cup B) = P(A) + P(B) - P(A \cap B)$$

Hence, the proof.

Corollary 1: If A, B, C are three events then

$$P(A \cup B \cup C) = P(A) + P(B) + P(C) - P(A \cap B) - P(B \cap C) - P(C \cap A) + P(A \cap B \cap C)$$

Corollary 2: If the events A, B, C are disjoint (independent) sets then the above results reduce to:

(i) $P(A \cup B) = P(A) + P(B)$

(ii) $P(A \cup B \cup C) = P(A) + P(B) + P(C)$

Corollary 3: If ϕ is an empty set then $P(\phi) = 0$.

Corollary 4: If $A \subset B$ then $P(A) < P(B)$ and $P(B - A) = P(B) - P(A)$.

Corollary 5: If A and B are any two events then

$$P(A) = P(A \cap B) + P(A \cap \bar{B})$$

$$\text{and } \quad P(B) = P(A \cap B) + P(\bar{A} \cap B)$$

Example 1: Find the probability that a leap year selected at random will contain 53 Sundays.

Solution: A leap year has 366 days

$$= 52 \text{ weeks} + 2 \text{ days}$$

For these two days we have the following possibilities:

(i) Mon, Tue (ii) Tue, Wed (iii) Wed, Thur (iv) Thur, Fri (v) Fri, Sat (vi) Sat, Sun (vii) Sun, Mon.

Here, $n = 7$ and $m = 2$

Hence, the probability that a leap year contains 53.

Sundays $= \frac{m}{n} = \frac{2}{7}.$

Example 2: Given A and B two events with $P(A \cup B) = \frac{7}{8}$, $P(A \cap B) = \frac{1}{4}$, and $P(A \cap \bar{B}) = \frac{5}{8}$. Find $P(A)$, $P(B)$ and $P(\bar{A} \cap B)$.

Solution: We have

(i)

$$P(A) = P(A \cap B) + P(A \cap \bar{B})$$

$$= \frac{1}{4} + \frac{5}{8} = \frac{7}{8}$$

(ii)

$$P(A \cup B) = P(A) + P(B) - P(A \cap B)$$

$$\frac{7}{8} = \frac{7}{8} + P(B) - \frac{1}{4}$$

$$\therefore \quad P(B) = \frac{1}{4}$$

(iii)

$$P(B) = P(A \cap B) + P(\bar{A} \cap B)$$

$$\frac{1}{4} = \frac{1}{4} + P(\bar{A} \cap B)$$

$$\therefore \quad P(\bar{A} \cap B) = 0$$

Example 3: A can hit a target 3 times in 5 shots, B 2 times in 5 shots and C 3 times in 4 shots. They fire a volley. What is the probability that: (i) two shots hit (ii) at least two shots hit?

Solution:

$$\text{Probability of } A \text{ hitting the target} = \frac{3}{5}$$

$$\text{Probability of } B \text{ hitting the target} = \frac{2}{5}$$

and
$$\text{Probability of } C \text{ hitting the target} = \frac{3}{4}$$

(i) In order that two shots may hit the target, we have

P_1 = Chance that A and B hit and C fails to hit

$$= P(A)P(B)(1 - P(C)) = \frac{3}{5} \times \frac{2}{5}\left(1 - \frac{3}{4}\right) = \frac{6}{100} = 0.06$$

P_2 = Chance that B and C hit and A fails to hit

$$= (1 - P(A))P(B)P(C) = \left(1 - \frac{3}{5}\right) \times \frac{2}{5} \times \frac{3}{4} = \frac{12}{100} = 0.12$$

P_3 = Chance that C and A hit and B fails to hit

$$= P(A)(1 - P(B))P(C) = \frac{3}{5}\left(1 - \frac{2}{5}\right)\frac{3}{4} = \frac{27}{100} = 0.27$$

Since these are mutually exclusive events, the probability that any two shots hit

$$= P_1 + P_2 + P_3 = \frac{6}{100} + \frac{12}{100} + \frac{27}{100} = \frac{45}{100} = 0.45$$

(ii) In order that at least two shots may hit the target we must also consider the case of all A, B, C hitting the target

i.e., P_4 = Chance that A, B, C all hit

$$= P(A)P(B)P(C) = \frac{3}{5} \times \frac{2}{5} \times \frac{3}{4} = \frac{18}{100} = 0.18$$

Since all these are mutually exclusive events, we have probability of at least shots hit

$$= P_1 + P_2 + P_3 + P_4 = \frac{6}{100} + \frac{12}{100} + \frac{27}{100} + \frac{18}{100} = \frac{63}{100} = 0.63$$

Example 4: Three machines A, B and C manufacture respectively 0.4, 0.5 and 0.1 of the total production. The percentage of defective items produced by A, B and C is 2, 4 and 1 per cent respectively. For an item chosen at random, what is the probability that it is defective?

Solution: The defective item produced by machine A

$$= .4 \times \frac{2}{100} = \frac{0.8}{100}$$

The defective item produced by machine B

$$= .5 \times \frac{4}{100} = \frac{2}{100}$$

and the defective item produced by machine C

$$= .1 \times \frac{1}{100} = \frac{0.1}{100}$$

∴ The total defective items produced by machines A, B and C

$$= \frac{.8}{100} + \frac{2}{100} + \frac{.1}{100} = \frac{2.9}{100}$$
$$= 0.029$$

Example 5: An urn contains 10 black and 10 white balls. Find the probability of drawing two balls of the same colour.

Solution: Given 10 black and 10 white balls.

∴ Probability of drawing two black balls $= {}^{10}C_2/{}^{20}C_2$
and probability of drawing two white balls $= {}^{10}C_2/{}^{20}C_2$

∴ Probability of drawing two balls of the same colour

$$= P(B \cup W) = P(B) + P(W)$$

$$= {}^{10}C_2/{}^{20}C_2 + {}^{10}C_2/{}^{20}C_2 = \frac{9}{19} = 0.4737$$

Example 6: In a given race, the odds is favour of horses H_1, H_2, H_3 and H_4 are 1 : 3, 1 : 4, 1 : 5 and 1 : 6 respectively. Find the probability that one of them wins the race.

Solution: Given $P(H_1)=\frac{1}{4}$, $P(H_2)=\frac{1}{5}$, $P(H_3)=\frac{1}{6}$ and $P(H_4)=\frac{1}{7}$.

$$\therefore \quad P(H_1 \text{ or } H_2 \text{ or } H_3 \text{ or } H_4) = P(H_1)+P(H_2)+P(H_3)+P(H_4)$$

$$= \frac{1}{4}+\frac{1}{5}+\frac{1}{6}+\frac{1}{7}$$

$$= \frac{319}{420} = 0.7595$$

6.1.4 Conditional Probability and Independent Events

Consider the difference between choosing two items at random, one after the other from a lot with and without replacing the first, before choosing the second. For example, consider the two events as drawing two cards A and B from a set of playing cards one after the other (i) with replacement $P(A)=P(B)=\frac{1}{52}$ (ii) without replacement. If we are choosing second card B without replacing the first then $P(A)=\frac{1}{52}$ and $P(B)=\frac{1}{51}$ as the total number of cards reduced to 51. That is to say that the occurrence of the first event has affected the occurrence of the second. Such events are said to be dependent events. We need the concept of conditional probability. If A and B are two events associated with an experiment E, we define $P(B/A)$ the conditional probability of the event B given that A has occurred.

$$\text{Thus} \quad P(B/A) = \frac{P(A\cap B)}{P(A)}$$

$$\text{or} \quad P(A\cap B) = P(A)\cdot P(B/A)$$

Similarly, the conditional probability of A given that B has occurred is given by

$$P(A/B) = \frac{P(A\cap B)}{P(B)}$$

$$\therefore \quad P(A\cap B) = P(B)P(A/B)$$

$$\therefore \quad P(A\cap B) = P(A)P(B/A) = P(B)P(A/B)$$

Corollary 1: If A, B, C are three events then

$$P(A\cap B\cap C) = P(A)P(B/A)P(C/A\cap B)$$

Corollary 2: If A and B are independent events, then

$$P(A \cap B) = P(A/B)P(B) = P(A) \cdot P(B)$$

$$P(A \cap B) = P(B/A)P(A) = P(B) \cdot P(A)$$

Definition: Two events are said to be independent events if occurrence of one does not affect the occurrence of the other event. Thus, if A and B are two independent events then

$$P(A \cap B) = P(A) \cdot P(B)$$

Multiplication Law for Mutually Independent Events

Theorem 1: If A and B are two independent events then prove that $P(A \cap B) = P(A) \cdot P(B)$.

Proof. Let A and B be two independent events. Let A happen in m_1 ways and fail n_1 ways.

$$\therefore \quad P(A) = \frac{m_1}{m_1 + n_1}$$

Also, let B happen in m_2 ways and fail n_2 ways.

$$\therefore \quad P(B) = \frac{m_2}{m_2 + n_2}$$

Now there are four possibilities:

(i) A and B both may happen, the number of ways $= m_1 m_2$

(ii) A may happen and B may fail, the number of ways $= m_1 n_2$

(iii) A may fail and B may happen, number of ways $= n_1 m_2$

(iv) A and B both may fail, number of ways $= n_1 n_2$

$\therefore$ Total number of ways

$$= m_1 m_2 + m_1 n_2 + n_1 m_2 + n_1 n_2$$

$$= m_1 (m_2 + n_2) + n_1 (m_2 + n_2)$$

$$= (m_1 + n_1)(m_2 + n_2)$$

Hence, the probability of the happening of both A and B

$$P(A \cap B) = \frac{m_1 m_2}{(m_1 + n_1)(m_2 + n_2)}$$

$$= \frac{m_1}{m_1 + n_1} \cdot \frac{m_2}{m_2 + n_2} = P(A) \cdot P(B)$$

$$\therefore \quad P(A \cap B) = P(A) \cdot P(B)$$

Theorem 2: If A and B are independent events then prove that (i) A and $\bar{B}$ (ii) $\bar{A}$ and B (iii) $\bar{A}$ and $\bar{B}$ are independent.

Proof. Since A and B are independent we have

$$P(A \cap B) = P(A) \cdot P(B)$$

(i)

$$P(A \cap \bar{B}) = P(A) - P(A \cap B)$$

$$= P(A) - P(A) \cdot P(B)$$

$$= P(A)(1 - P(B))$$

$$= P(A) \cdot P(\bar{B})$$

Hence, A and $\bar{B}$ are independent events.

(ii)

$$P(\bar{A} \cap B) = P(B) - P(A \cap B)$$

$$= P(B) - P(A)P(B)$$

$$= (1 - P(A)) \cdot P(B)$$

$$= P(\bar{A})P(B)$$

Hence, $\bar{A}$ and B are independent events.

(iii)

$$P(\bar{A} \cap \bar{B}) = P(\overline{A \cup B})$$

$$= 1 - P(A \cup B)$$

$$= 1 - [P(A) + P(B) - P(A \cap B)]$$

$$= 1 - [P(A) + P(B) - P(A) \cdot P(B)]$$

$$= (1 - P(A))(1 - P(B))$$

$$= P(\bar{A}) \cdot P(\bar{B})$$

Hence, $\bar{A}$ and $\bar{B}$ are independent events.

Theorem 3: Let $A_1, A_2, A_3, \ldots A_n$ be n independent events with respective probabilities of occurrence $P_1, P_2, P_3, \ldots P_n$ then the probability of occurrence of at least one of them is given by

$$P(A_1 \cup A_2 \cup A_3, \ldots \cup A_n) = 1 - [(1 - P_1)(1 - P_2) \ldots (1 - P_n)]$$

Proof. We know that

$$P(E) + P(\bar{E}) = 1$$

$$\therefore \quad P(E) = 1 - P(\bar{E})$$

Given $P(A_i) = P_i \quad i = 1, 2, 3, \ldots n$

$$P(\bar{A}_i) = 1 - P_i \quad i = 1, 2, 3, \ldots n$$

$$\therefore \quad P(A_1 \cup A_2 \cup A_3 \ldots \cup A_n)$$

$$= 1 - P(\overline{A_1 \cup A_2 \cup \ldots \cup A_n})$$

$$= 1 - p(\bar{A}_1 \cap \bar{A}_2 \cap \ldots \cap \bar{A}_n) \qquad \text{by De Morgan's laws}$$

$$= 1 - P(\bar{A}_1) P(\bar{A}_2) \ldots P(\bar{A}_n)$$

$$= 1 - (1 - P_1)(1 - P_2) \ldots (1 - P_n).$$

Hence the proof.

Example 1: Two persons A and B fire at a target independently and have a probability 0.6 and 0.7 respectively of hitting the target. Find the probability that the target is destroyed.

Solution: Probability that the target is destroyed

= Probability that at least one of A or B will hit the target.

$$= 1 - (1 - P(A))(1 - P(B))$$

$$= 1 - (1 - 0.6)(1 - 0.7)$$

$$= 0.88.$$

Example 2: Three students A, B and C write an entrance examination. Their chances of passing is respectively $\frac{1}{2}, \frac{1}{3}$ and $\frac{1}{4}$. Find the probability that (i) at least one of them will pass, (ii) all will pass (iii) at least two will pass.

Solution: Given data is $P(A) = \frac{1}{2}$, $P(B) = \frac{1}{3}$ and $P(C) = \frac{1}{4}$.

(i) Probability that at least one will pass

$$= 1 - (1 - P(A))(1 - P(B))1 - (1 - P(C))$$

$$= 1 - \left(1 - \frac{1}{2}\right)\left(1 - \frac{1}{3}\right)\left(1 - \frac{1}{4}\right)$$

$$= \frac{3}{4} = 0.75$$

(ii) Probability that all will pass

$$P(A \cap B \cap C) = P(A) \cdot P(B) \cdot P(C)$$

$$= \frac{1}{2} \cdot \frac{1}{3} \cdot \frac{1}{4}$$

$$= \frac{1}{24} = 0.0416$$

(iii) Probability that at least two will pass

$$= P(A \cap B \cap \bar{C}) + P(A \cap \bar{B} \cap C) + P(\bar{A} \cap B \cap C) + P(A \cap B \cap C)$$

$$= \frac{1}{2} \cdot \frac{1}{3}\left(1 - \frac{1}{4}\right) + \frac{1}{2} \cdot \left(1 - \frac{1}{3}\right)\frac{1}{4} + \left(1 - \frac{1}{2}\right)\frac{1}{3} \cdot \frac{1}{4} + \frac{1}{2} \cdot \frac{1}{3} \cdot \frac{1}{4}$$

$$= \frac{1}{8} + \frac{1}{12} + \frac{1}{24} + \frac{1}{24}$$

$$= \frac{7}{24} = 0.2916$$

Example 3: Anand and Ashok, appear in an interview for two vacancies in the same post. The probability of Anand's selection is $\frac{1}{7}$ and that of Ashok's selection is $\frac{1}{5}$ what is the probability that:

(i) both of them will be selected
(ii) only one of them will be selected and
(iii) none of them will be selected

Solution: Let us take Anand's selection as $P(A) = \frac{1}{7}$ and Ashok's selection as $P(B) = \frac{1}{5}$.

(i) Both of them will be selected

$$= P(A \cap B) = P(A)P(B) = \frac{1}{7} \cdot \frac{1}{5} = \frac{1}{35} = 0.0285$$

(ii) Probability of only one will be selected

$$= P(A \cap \overline{B}) + P(\overline{A} \cap B)$$

$$= \frac{1}{7} \cdot \left(1 - \frac{1}{5}\right) + \left(1 - \frac{1}{7}\right)\frac{1}{5}$$

$$= \frac{4}{35} + \frac{6}{35} = \frac{10}{35} = \frac{2}{5} = 0.4$$

(iii) Probability of none of them will be selected

$$= P(\overline{A} \cap \overline{B}) = P(\overline{A}) \cdot P(\overline{B})$$

$$= \frac{6}{7} \times \frac{4}{5} = \frac{24}{35} = 0.6857$$

Example 4: Three balls are drawn successively from a box containing 6 red, 4 white and 5 blue balls. Find the probability that they are drawn in order red, white and blue, if each ball is (i) replaced and (ii) not replaced.

Solution: Let A be the event red, B be the event white and C be the event blue.

(i) When each ball is replaced the events are independent

$$\therefore \quad P(A \cap B \cap C) = P(A)P(B)P(C)$$

$$= \frac{6}{15} \cdot \frac{4}{15} \cdot \frac{5}{15}$$

$$= \frac{8}{225} = 0.0355$$

(ii) When each ball is not replaced the events are dependent. Hence, by using conditional probability formula, we have

$$P(A \cap B \cap C) = P(A)P(B|A)P(C|A \cap B)$$

$$= \left(\frac{6}{15}\right)\left(\frac{4}{5+4+5}\right)\left(\frac{5}{5+3+5}\right)$$

$$= \frac{6}{15} \cdot \frac{4}{14} \cdot \frac{5}{13}$$

$$= \frac{8}{182} \cdot = 0.0439$$

Example 5: Bag *A* contains 2 white and 4 black balls. Another bag *B* contains 5 white and 7 black balls. A ball is transformed from the bag *A* to the bag *B*. Then a ball is drawn from the bag *B*. Find the probability that it is white.

Solution: The probability of drawing a white ball from bag *B* will depend on whether the transformed ball is white or black.

If the black ball is transformed, its probability is $\frac{4}{6}$. Then we have 5 white and 8 black balls in bag *B*.

$\therefore$ Probability of drawing white ball from bag *B* is $\frac{5}{13}$.

$\therefore$ Probability of drawing white ball from bag *B* when black ball is transformed from *A* to *B*

$$= \frac{4}{6} \times \frac{5}{13} = \frac{10}{39} = 0.2564$$

Similarly, probability of drawing a white ball from bag *B* when white ball is transformed from bag *A* to bag *B*

$$= \frac{2}{6} \times \frac{6}{13} = \frac{2}{13} = 0.1538$$

$\therefore$ The required probability $= \frac{10}{39} + \frac{2}{13} = \frac{16}{39} = 0.4102$

Example 6: Out of 20 mangoes 12 mangoes are found to be defective. If these mangoes are chosen one after the other at random what is the probability that:

(i) The first two mangoes are defective?

(ii) The first two mangoes are non-defective?

(iii) Among the first two mangoes there is one defective and one non-defective.

Solution: Let A be a defective mango and B be a non-defective mango.

Then $P(A) = \frac{12}{20}$, $P(B) = \frac{8}{20}$

(i) Probability that the first two mangoes are defective

$$= \frac{12}{20} \times \frac{11}{19} = \frac{33}{95} = 0.3473$$

(ii) Probability that the first two mangoes are non-defective

$$= \frac{8}{20} \times \frac{7}{19} = \frac{14}{95} = 0.1473$$

(iii) Probability that among the first two mangoes one is defective and one non-defective

= Probability of the first defective and the second non-defective

+ Probability of the first non-defective and the second defective

$$= \frac{12}{20} \times \frac{8}{19} + \frac{8}{20} \times \frac{12}{19}$$

$$= \frac{48}{95} \cdot = 0.5053$$

Example 7: A student takes his examination in four subjects A, B, C and D. He estimates his chances of passing A as $\frac{4}{5}$, B as $\frac{3}{4}$, C as $\frac{5}{6}$ and D as $\frac{2}{3}$. To qualify he must pass in A and at least two other subjects. What is the probability that he qualifies?

Solution: Given $P(A) = \frac{4}{5}$, $P(B) = \frac{3}{4}$, $P(C) = \frac{5}{6}$ and $P(D) = \frac{2}{3}$.

Probability of passing at least two subjects from B, C and D

$$= P(B)P(C)P(\overline{D}) + P(B)P(\overline{C})P(D) + P(\overline{B})P(C)P(D) + P(B)P(C)P(D)$$

$$= \left(\frac{3}{4}\right)\left(\frac{5}{6}\right)\left(1\frac{2}{3}\right) + \frac{3}{4}\left(1\frac{-5}{6}\right)\left(\frac{2}{3}\right) + \left(1\frac{-3}{4}\right)\left(\frac{5}{6}\right)\left(\frac{2}{3}\right) + \left(\frac{3}{4}\right)\left(\frac{5}{6}\right)\left(\frac{2}{3}\right)$$

$$= \frac{5}{24} + \frac{5}{36} + \frac{1}{12} + \frac{5}{12} = \frac{61}{72} = 0.8472$$

∴ Probability of passing A and at least two subjects

$$= \frac{4}{5} \times \frac{61}{72} = \frac{61}{90} = 0.677$$

Example 8: A bag contains 5 white and 3 black balls. Two balls are drawn at random one after the other without replacement. Find the probability that both balls drawn are black.

Solution: Probability of drawing a black ball in the first attempt is

$$P(A) = \frac{3}{5+3} = \frac{3}{8}$$

Probability of drawn the second black ball given that the first ball drawn is black

$$P(B|A) = \frac{2}{5+2} = \frac{2}{7}$$

∴ Probability that both the balls are black is given by

$$P(A \cap B) = P(A)P(B|A)$$

$$= \frac{3}{8} \times \frac{2}{7}$$

$$= \frac{3}{28} = 0.1071$$

6.2 BAYE'S THEOREM

Statement: If A_1, A_2, $A_3 \ldots A_n$ are disjoint sets whose union is the sample space 'S' and E is any event in S then

$$P(A_i|E) = \frac{P(A_i)\,P(E|A_i)}{\sum_{i=1}^{n} P(A_i)P(E|A_i)}$$

Proof. Given

$$S = A_1 \cup A_2 \cup A_3 \cdots \cup A_n$$

we have

$$E = E \cap S = E \cap (A_1 \cup A_2 \cup A_3 \cdots \cup A_n)$$

$$= (E \cap A_1) \cup (E \cap A_2) \cdots \cup (E \cap A_n)$$

where $E \cap A_1, E \cap A_2 \ldots E \cap A_n$ are disjoint.

Hence, by addition theorem

$$P(E) = P(E \cap A_1) + P(E \cap A_2) \ldots + P(E \cap A_n)$$

By using $P(A \cap B) = P(A) \cdot P(B|A)$ we get

$$P(E) = P(A_1)P(E|A_1) + P(A_2)P(E|A_2) + \ldots + P(A_n)P(E|A_n)$$

$$= \sum_{i=1}^{n} P(A_i)P(E|A_i)$$

Also $P(A_i|E) = \dfrac{P(A_i \cap E)}{P(E)}$

$$= \frac{P(A_i)\, P(E|A_i)}{P(E)}$$

$$\therefore \quad P(A_i|E) = \frac{P(A_i)\, P(E|A_i)}{\sum_{i=1}^{n} P(A_i)P(E|A_i)}$$

Hence the proof.

Example 1: An office has four secretaries handling respectively 20%, 60%, 15% and 5% of the files at all government reports. The probability that they misfile such reports are respectively 0.05, 0.1, 0.1 and 0.05. Find the probability that the misfiled report can be blamed on the first secretary.

Solution: Let S_1, S_2, S_3 and S_4 be the four secretaries and E be the event that the report is misfiled.

We have to find $P(S_1|E)$.

Given $P(S_1) = 0.2$, $P(S_2) = 0.6$, $P(S_3) = 0.15$ and $P(S_4) = 0.05$ And, $P(E|S_1) = 0.05$, $P(E|S_2) = 0.1$, $P(E|S_3) = 0.1$ and $P(E|S_4) = 0.05$.

$\therefore$ By Baye's theorem

$$P(S_1|E) = \frac{P(S_1)P(E|S_1)}{\sum_{i=1}^{n} P(S_i)P(E|S_i)}$$

$$= \frac{(0.2)(0.05)}{(0.2)(0.05)+(0.6)(0.1)+(0.15)(0.1)+(0.05)(0.05)}$$

$$= \frac{(0.2)(0.05)}{0.0875} = 0.114.$$

Example 2: Three machines M_1, M_2, M_3 produce respectively 60%, 30% and 10% of the total number of items of a factory. The defective percentage of output of these machines are respectively 2%, 3% and 4%. An item is selected at random and is found to be defective. Find the probability that the item was produced by machine A. Also find the probability that an item drawn is defective.

Solution: Given $M_1 = 0.6$, $P(M_2) = 0.3$, and $P(M_3) = 0.1$ Let E be the event that an item is defective. Then $P(E|M_1) = 0.02$, $P(E|M_2) = 0.03$ and $P(E|M_3) = 0.04$ we have to find the probability that defective item selected is produced by machine M_1.

$\therefore$ By Baye's theorem

$$P(M_1|E) = \frac{P(M_1)P(E|M_1)}{P(M_1)P(E|M_2)+P(M_2)P(E|M_2)+P(M_3)P(E|M_3)}$$

$$= \frac{0.6 \times 0.02}{0.6 \times 0.02 + 0.3 \times 0.03 + 0.1 \times 0.04}$$

$$= \frac{12}{25} = 0.48$$

And, probability that an item selected is defective

$$P(E) = P(M_1)P(E|M_1) + P(M_2)P(E|M_2) + P(M_3)P(E|M_3)$$

$$= 0.6 \times 0.02 + 0.3 \times 0.03 + 0.1 \times 0.04$$

$$= 0.025$$

Example 3: Three players A, B, C of a cricket team try for the captainship and their chances of getting it are in the proportion 4:2:3. The probability that A if is made the captain will drop a particular player D from the team is 0.3. The probability that B and C doing the same are 0.5 and 0.8 respectively. What is the probability that D will not be dropped from the team if one among A, B, C is made captain?

Solution: Given $P(A) = \frac{4}{9}$, $P(B) = \frac{2}{9}$ and $P(C) = \frac{3}{9}$.

Let E be the event that player D is dropped from the team. Then

$$P(E|A) = 0.3,\ P(E|B) = 0.5 \quad \text{and} \quad P(E|C) = 0.8$$

Hence, the probability of dropping the player D from the team

$$P(E) = P(A)P(E|A) + P(B)P(E|B) + P(C)P(E|C)$$

$$= 0.3 \times \frac{4}{9} + 0.5 \times \frac{2}{9} + 0.8 \times \frac{3}{9}$$

$$= \frac{4.6}{9} = 0.51$$

∴ The probability of not dropping the player D from the team

$$= 1 - P(E)$$

$$= 1 - 0.51 = 0.49$$

Example 4: The probability of conducting an examination on time is 0.95, if there is no strike by students and 0.25 if there is a strike. If the probability that there will be a strike is 0.65. Find the probability of holding the examination on time.

Solution: Given $P(A) = 0.65$ and $P(B) = 1 - P(A) = 0.35$

where A and B are respectively events that there is a strike by the students and there is no strike.

Let E be the event that the examination is held on time. Then we have to find $P(E)$

$$P(E|A) = 0.25 \quad \text{and} \quad P(E|B) = 0.95$$

∴ By Baye's theorem

$$P(E) = P(E|A)P(A) + P(E|B)P(B)$$

$$= (0.25)(0.65) + (0.95)(0.35)$$

$$= 0.495$$

Example 5: In a certain college 4% of boys and 1% of girls are taller than 1.8 m. Further more 60% of the students are girls. If a student is selected at random and is found taller than 1.8 m, what is the probability that the student is a girl?

Solution: Let G and B be girls and boys in the college respectively

$$\therefore \quad P(G) = \text{Probability of choosing a girl from all students}$$

$$= 60\% = 0.6$$

$$P(B) = \text{Probability of choosing a boy from all students}$$

$$= 40\% = 0.4.$$

Now,

$$P(E/G) = \text{Probability of choosing a student who is taller than 1.8 m from girls}$$

$$= 1\% = 0.01$$

$$P(E/B) = \text{Probability of choosing a student who is taller than 1.8 m from boys}$$

$$= 4\% = 0.04$$

$$\therefore \quad P(G/E) = \frac{P(G)P(E/G)}{P(G)P(E/G) + P(B)P(E/B)}$$

$$= \frac{0.6 \times 0.01}{0.6 \times 0.01 + (0.4)(0.04)}$$

$$= 0.2727$$

Example 6: The chance that doctor A will diagnose a disease correctly is 60%. The chance that a patient of A will die after correct diagnosis is 40% and the chance of death by wrong diagnosis is 70%. If a patient of A dies what is the chance that his disease was correctly diagnosed?

Solution: Define the events

X is the event that disease is correctly diagnosed.

Y is the event that disease is wrongly diagnosed.

Then $P(X) = 60\% = 0.6, \quad P(Y) = 40\% = 0.4$

Let E be the event that the patient of A dies.

Then $P(E/X) = 40\% = 0.4$

$P(E/Y) = 70\% = 0.7$

We have to find $P(X/E)$

$\therefore$ By Baye's theorem

$$P(X/E) = \frac{P(E/X)P(X)}{P(E/X)P(X) + P(E/Y)P(Y)}$$

$$= \frac{0.4 \times 0.6}{0.4 \times 0.6 + 0.7 \times 0.4}$$

$$= \frac{6}{13} = 0.4615$$

Example 7: Three bags A, B and C contain respectively 6 green and 4 red balls, 2 green and 6 red balls and 1 green and 8 red balls. 2 balls are drawn from one of the bags. If both the balls are found to be red, what is the probability that it came from bag B.

Solution: There are three bags A, B and C

$$\therefore \quad P(A) = P(B) = P(C) = \frac{1}{3}$$

Let E be the selection of 2 red balls and we have to find $P(B/E)$

$$P(E/A) = P(2 \text{ red balls from } A)$$

$$= \frac{4C_2}{10C_2} = \frac{2}{15}$$

$$P(E/B) = P(2 \text{ red balls from } B)$$

$$= \frac{6C_2}{8C_2} = \frac{15}{28}$$

$$P(E/C) = P(2 \text{ red balls from } C)$$

$$= \frac{8C_2}{9C_2} = \frac{7}{9}$$

∴ By Baye's theorem

$$P(B/E) = \frac{P(B)P(E/B)}{P(A)P(E/A) + P(B)P(E/B) + P(C)P(E/C)}$$

$$= \frac{\left(\frac{1}{3}\right)\left(\frac{15}{28}\right)}{\left(\frac{1}{3}\right)\left(\frac{2}{15}\right) + \left(\frac{1}{3}\right)\left(\frac{15}{28}\right) + \left(\frac{1}{3}\right)\left(\frac{7}{9}\right)}$$

$$= \frac{675}{1823} = 0.3703$$

Example 8: From 6 positive and 8 negative numbers 4 are chosen at random (without replacement) and multiplied. What is the probability that the product is a positive number.

Solution: To get the product of 4 numbers as positive, we should have $4P, ON$ or $2P, 2N$ or $OP\ 4N$
where P denotes a + ve number and N denotes a – ve number.
If E be the event that the product of 4 numbers is + ve then

$$P(E) = \frac{{}^6C_4 \times {}^8C_0}{{}^{14}C_4} + \frac{{}^6C_2 \times {}^8C_2}{{}^{14}C_4} + \frac{{}^6C_0 \times {}^8C_4}{{}^{14}C_4}$$

$$= \frac{15 + 420 + 70}{1001} = \frac{505}{1001} = 0.5045$$

Example 9: Two persons A and B toss an unbiased coin alternatively on the understanding that the first who gets the head wins. If A starts the game, find their respective chance of winning.

Solution: Probability of getting a head $= \frac{1}{2}$

Then A can win in $1^{st}, 3^{rd}, 5^{th} \ldots$ throws.

∴ The chance of A's winning $= \frac{1}{2} + \left(\frac{1}{2}\right)^2 \frac{1}{2} + \left(\frac{1}{2}\right)^4 \frac{1}{2} + \left(\frac{1}{6}\right)^2 \frac{1}{2} + \ldots$

$$= \frac{\frac{1}{2}}{1 - \left(\frac{1}{2}\right)^2} = \frac{1}{2} \cdot \frac{4}{3} = \frac{2}{3}$$

Hence, the chance of B's winning $= 1 - \frac{2}{3} = \frac{1}{3} = 0.33$

Example 10: In a bolt factory, there are four machines M_1, M_2, M_3 and M_4 manufacturing 20%, 15%, 25% and 40% of the total output respectively of their outs 5%, 4%, 3% and 2% in the same order are defective bolts. A bolt is chosen at random from the factory's production and is found defective, what is the probability that the bolt was manufactured by machine M_1 or machine M_4?

Ans: 0.3175, 0.254

Example 11: The contents of three urns are: 1 white, 2 red, 3 green balls; 2 white, 1 red, 1 green balls and 4 white, 5 red, 3 green random. These are found to be one white and one green. Find the probability that the balls so drawn came from the third urn.

Ans: $\frac{15}{59} = 0.2542$

Example 12: There are three bags: first bag containing 1 white, 2 red, 3 green balls; second 2 white, 3 red, 1 green balls and third 3 white, 1 red, 2 green balls. Two balls are drawn from a bag chosen at random. These are found to be one white and one red. Find the probability that the balls so drawn came from the second bag.

Ans: $\frac{6}{11} = 0.5454.$

Hint: $P(B_1) = P(B_2) = P(B_3) = \frac{1}{3}$

QUESTION BANK

1. A problem in mathematics is given to three students A, B, C whose chances of solving it are $\frac{1}{4}, \frac{1}{5}$ and $\frac{1}{7}$ respectively. What is the probability that the problem will be solved.
2. Kiran can solve 30% of the problem in a text, Pooja can solve 40% and Rupa can solve 50% of them. If a randomly selected problem is given to them, what is the probability that it is solved.
3. A company producing transistor radios has three assembly plants producing 15%, 35% and 50% respectively of its outputs. Suppose that the probabilities that a radio produced by these plants is defective are 0.01, 0.05 and 0.02 respectively. A radio is chosen at random and found

to be defective, what is the probability that it came from the first plant or the second plant?

4. An assembly plant purchases 50%, 30% and 10% of its voltage stabilizers respectively from firms A, B and C only. 90% 75% and 50% of the stabilizers purchased from firms A, B and C work according to specifications. A voltage stabilizer is selected at random from the assembly plant, what is the probability that it works according to specifications? What is the probability that a voltage stabilizers working according to specification came from B?

UNIT 7

Probability Theory – II

7.1 RANDOM VARIABLES

Definition: A random variable is a function which takes real values which are determined by the outcomes of the random experiment.

For example, consider a random experiment of three coins tossed. Then the sample space 'S' consists of $2^3 = 8$ points as given below:

$$S = \{HHH, HHT, HTH, THH, HTT, THT, TTH, TTT\}$$

Let the number of heads obtained be X. Then X can take any one of the values 0, 1, 2.

Outcome:	HHH	HHT	HTH	THH	HTT	THT	TTH	TTT
values of X:	3	2	2	2	1	1	1	0

If the sample points are denoted by w_1, w_2, $w_3 \ldots w_8$ then to each outcome w of the random experiment, we have

$$X = X(w) \quad \text{i.e., } X(w_1) = 3,\ X(w_2) = 2\ldots X(8) = 0$$

Thus, random variable may be defined as a real valued function on the sample space, taking values on the real line R $(-\infty, \infty)$. In other words, random variable is a function which takes real values which are determined by the outcomes of the random experiment.

7.1.1 Discrete and Continuous Random Variables

If the random variable X assumes only a finite or countably infinite set of values, it is known as discrete random variable.

If the random variable X can assume infinite and uncountable set of values, it is said to be a continuous random variable.

For example,

Marks obtained by students in a test. The number of students in a college, the number of defective mangoes in a basket of mangoes. The number of accidents taking place on a busy road etc., are all discrete random variables. And, the age, height, weight of students in a class are all continuous random variables.

7.1.2 Probability Distribution of a Discrete Random Variable

Let X be a discrete random variable assuming the values $x_1, x_2, x_3 \ldots x_n$ with each possible outcome x_i, we associate a number $P_i = P(X = x_i) = P(x_i)$ is called the probability mass function of the random variable X if the following conditions are satisfied.

(i) $P(x_i) \geq 0$ for all i

(ii) $\sum\limits_{i=1}^{n} P(x_i) = 1$

i.e., $P(x_1) + P(x_2) + \ldots + P(x_n) = 1$

The set of all possible ordered pairs $\{x_i, p(x_i)\}$ is called the probability distribution of the random variable X.

Consider the experiment of tossing of two fair coins. If the random variable X denotes the number of heads then X can take the values $0, 1, 2$ with probabilities $\frac{1}{4}, \frac{1}{2}, \frac{1}{4}$ respectively. Thus, the probability distribution for the number of heads is given by

X:	0	1	2	T
$P(X)$:	$\frac{1}{4}$	$\frac{1}{2}$	$\frac{1}{4}$	$= 1$

From the above table, we have

$$P(x_i) \geq 0 \quad i = 1, 2, 3 \quad \text{and} \quad \sum_{i=1}^{n} P(x_i) = 1$$

7.1.3 Probability Distribution of a Continuous Random Variable

Let X be a continuous random variable and $f(x)$ be any continuous function of x. Then $f(x)\, dx$ represents the probability that x falls in the interval $\left[x - \frac{dx}{2}, x + \frac{dx}{2}\right]$ and we write symbolically as

$$P\left[x-\frac{dx}{2}, x+\frac{dx}{2}\right] = P\left[x-\frac{dx}{2} \le X \le x+\frac{dx}{2}\right]$$

$$= f_X(x)\, dx$$

Also $f(x)\, dx$ represents the area bounded by the curve $y = f(x)$, x-axis and the ordinates at the points $x - \frac{dx}{2}$ and $x + \frac{dx}{2}$. The function $f(x)\, dx$ is called the probability density function (*P.D.F.*). The curve $y = f(x)$ is called the probability curve.

In short, $f(x)$ is said to be a probability density function if:

(i) $f(x) \ge 0$ for all $x \in R$

(ii) $\int_{-\infty}^{\infty} f(x)\, dx = 1$ and

(iii) The probability for a variate value in the finite interval $[a, b]$ is

$$\int_a^b f(x)\, dx = P[a \le x \le b]$$

7.1.4 Cumulative Distribution Function (*CDF*) or Distribution Function

Let X be a random variable, discrete or continuous, we define F to be the cumulative distribution function of the random variable where $F(x) = P(X \le x)$

$$F(x_i) = P(X \le x_i)$$

$$= P(x_1) + P(x_2) + \dots + P(x_i)$$

Also

$$P(X > x_i) = P(x_{i+1}) + P(x_{i+2}) + \dots + P(x_n)$$

If X is a continuous random variable with probability density function then

$$F(x) = \int_{-\infty}^{\infty} P(x)\, dx$$

Also

$$P(x) = \frac{d}{dx}(F(x))$$

If a and b are two specified values of X with $b > a$ we have

$$P(a \le X \le b) = P(X \le b) - P(X > a)$$

Example 1: A die is tossed twice. Getting an odd number is termed success. Find the probability distribution of the number of successes.

Solution: There are in all 3 cases to odd numbers 1,3,5

$\therefore$ Probability of success $= \frac{3}{6} = \frac{1}{2} = P$

and probability of failure $= 1 - P = 1 - \frac{1}{2} = \frac{1}{2} = q$

If X denotes the number of successes in two throws of a die then X is a random variable which takes the values $0, 1, 2$.

$$P(X = 0) = P(F.F.) = \frac{1}{2} \times \frac{1}{2} = \frac{1}{4}$$

$$P(X = 1) = P(S.F.) + P(FS) = \frac{1}{2} \times \frac{1}{2} + \frac{1}{2} \times \frac{1}{2} = \frac{1}{2}$$

$$P(X = 2) = P(S.S.) = \frac{1}{2} \times \frac{1}{2} = \frac{1}{4}$$

Hence, the probability distribution of X is given by

X:	0	1	2
$P(X)$:	$\frac{1}{4}$	$\frac{1}{2}$	$\frac{1}{4}$

$\therefore$ $P(x_i) \ge 0$

and $\sum P(x_i) = \frac{1}{4} + \frac{1}{2} + \frac{1}{4} = 1$

Example 2: Obtain probability distribution of X for the number of heads in 3 tosses of a coin.

Solution: When we toss three coins then the sample space is

$$S = \{HHH,\ HHT,\ THH,\ HTH,\ HTT,\ THT,\ TTH,\ TTT\}$$

Then define the random variable X as the number of heads obtained in three tosses.

$\therefore$ $X = \{3, 2, 1, 0\}$

$A_0 = \{TTT\}$ $A_1 = \{HTT, THT, TTH\}$ $A_2 = \{HHT, HTH, THH\}$

$A_3 = \{HHH\}$

Clearly $A_0 \cup A_1 \cup A_2 \cup A_3 = S$

$$\text{and} \quad P(0) = P(A_0) = \frac{1}{2} \times \frac{1}{2} \times \frac{1}{2} = \frac{1}{8}$$

$$P(1) = P(A_1) = \frac{3}{8}$$

$$P(2) = P(A_2) = \frac{3}{8}$$

$$P(3) = \frac{1}{8}$$

Hence, the probability distribution is given by

X:	0	1	2	3
$P(X)$:	$\frac{1}{8}$	$\frac{3}{8}$	$\frac{3}{8}$	$\frac{1}{8}$

clearly, $p(x) \geq 0$

and $\sum p(x_i) = 1$

7.2 MATHEMATICAL EXPECTATION

Let X be the discrete random variable, assuming the values $x_1, x_2, x_3 \ldots x_n$ with the probabilities of occurrence $P_1, P_2, P_3 \ldots P_n$ respectively, then the mathematical expectation or just expectation of X denoted by $E(X)$ is defined as

$$E(X) = x_1 p_1 + x_2 p_2 + x_3 p_3 + \ldots + x_n p_n$$

$$= \sum_{i=1}^{n} x_i \, p_i$$

i.e., mean of the distribution

$$\therefore \quad \text{mean} = \bar{x} = E(X) = \sum_{i=1}^{n} x_i \, p_i$$

(i) For a continuous random variable X, we define $E(X)$ as

$$E(X) = \int_{-\infty}^{\infty} x f(x) \, dx = \mu$$

$$= \text{mean}$$

where $f(x)$ is the probability density function.

(ii) If X is a discrete random variable then the variance of X denoted by var(X), is defined as

$$\text{var}(X) = E(X^2) - \lfloor E(X) \rfloor^2$$

(iii) If X is a continuous random variable then the var(X) is given by

$$\text{var}(X) = \int_{-\infty}^{\infty} x^2 f(x)\, dx - \left[\int_{-\infty}^{\infty} x\, f(x)\, dx\right]^2$$

and standard deviation $= \sigma = \sqrt{\text{var}(X)}$

Example 1: What is the expected number of heads appearing when a fair coin is tossed three times.

Solution: The probability distribution of X is

X:	0	1	2	3
$P(X)$:	$\frac{1}{8}$	$\frac{3}{8}$	$\frac{3}{8}$	$\frac{1}{8}$

$$\therefore \quad E(X) = \sum x_i\, p(x_i)$$

$$= 0 \times \frac{1}{8} + 1 \times \frac{3}{8} + 2 \times \frac{3}{8} + 3 \times \frac{1}{8}$$

$$= \frac{12}{8} = \frac{3}{2} = 1.5$$

Example 2: Find the mean and variance of the probability distribution given by

X:	1	2	3	4	5
$P(x)$:	0.2	0.35	0.25	0.15	0.05

Solution: We have

$$\text{Mean} = E(X) = \sum_{i=1}^{5} x_i\, p(x_i)$$

$$= 1 \times 0.2 + 2 \times 0.35 + 3 \times 0.25 + 4 \times 0.15 + 5 \times 0.05$$

$$= 2.5$$

$$\text{Var}(X) = \sigma^2 = E(X^2) - \lfloor E(X) \rfloor^2$$

$$= 1^2 \times 0.2 + 2^2 \times 0.35 + 3^2 \times 0.25 + 4^2 \times 0.15 + 5^2 \times 0.05 - (2.5)^2$$

$$= 1.25$$

Example 3: The probability distribution of a finite random variable X is given by

X:	-2	-1	0	1	2	3
$P(x)$:	0.1	k	0.2	$2k$	0.3	k

Find the value of K. Also find mean and variance of X.

Solution: Since it is a probability distribution we have

$$\sum p(x_i) = 1$$

$$\therefore \quad 0.1 + K + 0.2 + 2k + 0.3 + K = 1$$

$$\therefore \quad K = 0.1$$

$$\text{Mean} = E(X) = \sum x_i p(x_i)$$

$$= -2 \times 0.1 + (-1) \times 0.1 + 0 \times 0.2 + 1 \times 0.2 + 2 \times 0.3 + 3 \times 0.1$$

$$= 0.8$$

$$\text{Var}(X) = \sigma^2 = E(X^2) - \lfloor E(X) \rfloor^2$$

$$= (-2)^2 \times 0.1 + (-1)^2 \times 0.1 + 0^2 \times 0.2 + 1^2 \times 0.2 + 2^2 \times 0.3$$

$$+ 3^2 \times 0.1 - (0.8)^2$$

$$= 2.16$$

(i) $P(X < 1) = P(-2) + P(-1) + P(0)$

$$= .1 + k + .2 = 0.4$$

(ii) $P(-1 < X \leq 2) = P(0) + P(1) + P(2)$

$$= 0.2 + 0.2 + 0.3$$

$$= 0.7$$

Example 4: Find $E(X)$, $E(X^2)$ and σ^2 for the probability function $p(x)$ defined by

$$x_i = 1 \quad 2 \quad 3 \quad 4 \ldots n$$

$$p(x_i) = K \quad 2k \quad 3k \quad 4k \ldots nK$$

where K is a approximate constant.

Solution: We have $\sum p(x_i) = 1$

$$\therefore \quad P(x_1) + P(x_2) + \ldots + P(x_n) = 1$$

$$K + 2k + 3k + \ldots + k_n = 1$$

$$K(1 + 2 + 3 + \ldots + n) = 1$$

$$\therefore \quad K\frac{n(n+1)}{2} = 1 \qquad \therefore \quad K = \frac{2}{n(n+1)}$$

Now, $E(X) = \sum x_i P(x_i)$

$$= x_1 P(x_1) + x_2 P(x_2) + \ldots + x_n P(x_n)$$

$$= K\left(1^2 + 2^2 + 3^2 + \ldots + n^2\right)$$

$$= \frac{2}{n(n+1)} \frac{n(n+1)(2n+1)}{6}$$

$$= \frac{2n+1}{3}$$

Also $E(X^2) = K\left(1^3 + 2^3 + 3^3 + \ldots + n^3\right)$

$$= \frac{2}{n(n+1)} \frac{n^2(n+1)^2}{4}$$

$$= \frac{n(n+1)}{2}$$

and $\sigma^2 = E(X^2) - \lfloor E(X) \rfloor^2$

$$= \frac{n(n+1)}{2} - \frac{(2n+1)^2}{3^2}$$

$$= \frac{n^2 + n - 2}{18}$$

Example 5: The frequency of accidents per shift in a factory is shown in the following table. Determine mean and standard deviation.

Accidents per shift:	0	1	2	3	4
Frequency:	192	100	24	3	1

Solution: Total frequency = 320.

$\therefore$ Probability distribution for the number of accidents

x:	0	1	2	3	4
$p(x)$:	$\frac{192}{320}$	$\frac{100}{320}$	$\frac{24}{320}$	$\frac{3}{320}$	$\frac{1}{320}$

$$\therefore \quad \text{Mean} = E(X) = \sum_{i=0}^{4} x_i p(x_i)$$

$$= 0 \times \frac{192}{320} + 1 \times \frac{100}{320} + 2 \times \frac{24}{320} + 3 \times \frac{3}{320} + 4 \times \frac{1}{320}$$

$$= \frac{161}{320} = 0.5$$

And $\operatorname{var}(X) = E(X^2) - \lfloor E(X) \rfloor^2$

$$= 0^2 \times \frac{192}{320} + 1^2 \times \frac{100}{320} + 2^2 \times \frac{24}{320} + 3^2 \times \frac{3}{320}$$

$$+ 4^2 \times \frac{1}{320} - \left(\frac{121}{320}\right)^2$$

$$\sigma^2 = \frac{239}{320} - \left(\frac{161}{320}\right)^2$$

$$\therefore \quad \sigma = \sqrt{\frac{239}{320} - \left(\frac{161}{320}\right)^2}$$

$$= 0.703$$

Example 6: The probability distribution is given by the following table

x_i:	0	1	2	3	4	5	6
$p(x_i)$:	k	$3k$	$5k$	$7k$	$9k$	$11k$	$13k$

For what values of k this represents a valid probability distribution?

Determine $P(X < 4)$, $P(X \geq 5)$, $P(3 < X \leq 6)$ and $P(X \leq 2) > 3$

Solution: We have $\sum p(x_i) = 1$

$$\therefore \quad k + 3k + 5k + 7k + 9k + 11k + 13k = 1$$

$$\therefore \quad 49k = 1 \quad \therefore \quad k = \frac{1}{49}$$

(i) $P(X < 4) = P(0) + P(1) + P(2) + P(3)$

$$= k + 3k + 5k + 7k$$

$$= 16k = \frac{16}{49} = 0.3265$$

(ii) $P(X \geq 5) = P(5) + P(6)$

$$= 11k + 13k = 24k$$

$$= \frac{24}{49} = 0.4898$$

(iii) $P(3 < X \leq 6) = P(4) + P(5) + P(6)$

$$= 9k + 11k + 13k$$

$$= 33k = \frac{33}{49} = 0.6734$$

(iv) $P(X \leq 2) > .3 = P(0) + P(1) + P(2)$

$$= k + 3k + 5k = 9k > 0.3$$

$$= 9k > .3 \text{ or } k > \frac{1}{30}$$

Example 7: A die is tossed twice. Getting a number greater than 4 is considered a success. Find the mean and variance of the probability distribution of the number of successes.

Solution: Since the favourable cases for getting a number greater than 4 in a throw of a die are 5 and 6

i.e., 2. $\therefore$ Probability of success $= P = \frac{2}{6} = \frac{1}{3}$

and probability of failure $= q = 1 - \frac{1}{3} = \frac{2}{3}$

Let X denote the number of successes. Then X is a random variable taking the values 0, 1 and 2.

$$\therefore \quad P(X=0) = P(q)P(q) = \frac{2}{3} \times \frac{2}{3} = \frac{4}{9}$$

$$P(X=1) = P(p)P(q) + P(q)P(p) = \frac{1}{3} \times \frac{2}{3} + \frac{2}{3} \times \frac{1}{3} = \frac{4}{9}$$

$$P(X=2) = P(p)P(p) = \frac{1}{3} \times \frac{1}{3} = \frac{1}{9}$$

$$\therefore \quad \text{Mean} = E(X) = \sum XP(x) = 0 \times \frac{4}{9} + 1 \times \frac{4}{9} + 2 \times \frac{1}{9} = \frac{6}{9}$$

$$E(X^2) = \sum X^2 P(x) = 0 + \frac{4}{9} + \frac{4}{9} = \frac{8}{9}$$

$$\therefore \quad \text{Var}(X) = E(X^2) - \lfloor E(X) \rfloor^2$$

$$= \frac{8}{9} - \left(\frac{6}{9}\right)^2 = \frac{72-36}{81} = \frac{36}{81} = 0.444 = \sigma^2$$

Example 8: A continuous random variable X has the probability density function given by

$$f(x) = \begin{cases} 2e^{-2x} & x > 0 \\ 0, & \text{otherwise} \end{cases}$$

Find the expected value and the variance of X.

Solution: Expected value i.e., mean is given by

$$E(X) = \int_{-\infty}^{\infty} x\, f(x)\, dx$$

$$= \int_{0}^{\infty} 2xe^{-2x}\, dx$$

$$= \left[2x\frac{e^{-2x}}{-2} - 2\frac{e^{-2x}}{4}\right]_0^{\infty} = 0 - 0 + \frac{2}{4} = \frac{1}{2}$$

and $$E(X^2) = \int_0^{\infty} 2x^2 e^{-2x}\, dx$$

$$= \left[2x^2\frac{e^{-2x}}{-2} - 4x\frac{e^{-2x}}{4} + 4\frac{e^{-2x}}{-8}\right]_0^{\infty}$$

$$= 0 - \left(\frac{4}{-8}\right) = \frac{1}{2}$$

$$\therefore \quad \text{Var}(X) = E(X^2) - [E(X)]^2$$

$$= \frac{1}{2} - \frac{1}{4} = \frac{1}{4}$$

Example 9: The life in hours of an electronic device in 100 hours is a random variable X having the density function

$$f(x) = \begin{cases} 0, & x < 2 \\ \frac{1}{18}(3 + 2x), & 2 \le x \le 4 \\ 0, & x > 4 \end{cases}$$

Find the mean and variance of X.

Solution: The mean or expected value of X is given by

$$E(X) = \int_{-\infty}^{\infty} x f(x)\, dx$$

$$= \int_2^4 x \cdot \frac{1}{18}(3 + 2x)\, dx$$

$$= \frac{1}{18}\int_2^4 \left(3x + 2x^2\right) dx$$

$$= \frac{1}{18}\left[3\frac{x^2}{2} + 2\frac{x^3}{3}\right]_2^4$$

$$= \frac{156}{54}$$

$$\text{And } E(X^2) = \int_{-\infty}^{\infty} x^2 f(x)\, dx$$

$$= \frac{1}{18}\int_2^4 x^2(3+2x)\, dx$$

$$= \frac{1}{18}\left[3\cdot\frac{x^3}{3} + 2\frac{x^4}{4}\right]_2^4 = \frac{176}{18}$$

$$\therefore \quad \text{Var}(X) = E(X^2) - \lfloor E(X) \rfloor^2$$

$$= \frac{176}{18} - \left(\frac{156}{54}\right)^2$$

$$\therefore \quad \sigma^2 = 1.43$$

Example 10: Is the function defined as follows a density function?

$$f(x) = e^{-x} \qquad x \geq 0$$

$$= 0 \qquad x < 0$$

If so determine the probability that the variate having this density will fall in (1, 2).

Solution: (i) $f(x)$ is clearly ≥ 0 for every $x \in (1, 2)$

$$\text{and } \int_{-\infty}^{\infty} f(x)dx = \int_{-\infty}^{0} 0dx + \int_{0}^{\infty} e^{-x}dx = 1$$

Hence the function $f(x)$ satisfies the condition for a density function.

(ii) The probability that the variate having this density will fall in (1, 2) is

$$= P(1 \leq x \leq 2) = \int_1^2 e^{-x}dx$$

$$= e^{-1} - e^{-2} = .368 - .138 = .233$$

$\therefore$ Probability will fall in the interval (1, 2).

Example 11: The waiting time (in minutes) in between two successive speeders spotted by a radar is a continuous random variable with distribution function

$$F(x) = \begin{cases} 0 & x \le 0 \\ 1 - e^{-8x} & x > 0 \end{cases}$$

Find the probability of waiting less than 12 minutes between successive speeders: (i) using *CDF* and (ii) *PDF*.

Solution: Using the given *CDF* we find the required probability as

$$P(x \le 12) = F(12) = 1 - e^{-9.6} \tag{1}$$

Next we find that the *PDF*

$$P(x) = F'(x) = \begin{cases} 0, & x \le 0 \\ 8e^{-8x}, & x > 0 \end{cases}$$

Then we find

$$P(x \le 12) = \int_{-\infty}^{12} P(x)dx = \int_{0}^{12} 8e^{-8x}dx$$

$$= 1 - e^{-9.6} \tag{2}$$

$\therefore$ (1) and (2) are identical.

Example 12: A random variable x has the density function

$$P(x) = \begin{cases} kx^2, & -3 \le x \le 3 \\ 0, & \text{elsewhere} \end{cases}$$

Evaluate k and find: (i) $P(1 \le x \le 2)$ (ii) $P(x \le 2)$ and $P(x > 1)$

Solution: In order that $P(x)$ may be probability density function, the two conditions to be satisfied are $P(x) \ge 0$ and $\int_{-\infty}^{\infty} P(x)dx = 1$

The given function satisfied the first condition if $k \ge 0$

The second condition is satisfied if

$$\int_{-\infty}^{\infty} P(x)dx = 1 \quad \text{i.e., if} \int_{-3}^{3} kx^2 dx = 1 \quad \therefore \quad k = \frac{1}{18}$$

Then

(i) $P(1 \le x \le 2) = \int_1^2 P(x)dx = \frac{1}{18}\int_1^2 x^2 dx = \frac{7}{54}$

(ii) $P(x \le 2) = \int_{-\infty}^{2} P(x)dx = \frac{1}{18}\int_{-3}^{2} x^2 dx = \frac{35}{54}$

(iii) $P(x > 1) = \int_1^{\infty} P(x)dx = \frac{1}{18}\int_1^3 x^2 dx = \frac{13}{27}$

Example 13: The probability density function of a continuously random variable is

$$P(x) = \begin{cases} kx(1-x)e^x, & 0 \le x \le 1 \\ 0, & \text{elsewhere} \end{cases}$$

Find k and evaluate mean and standard deviation.

Solution: Clearly $P(x) \ge 0$ and $\int_{-\infty}^{\infty} P(x)dx = 1$

$$\text{i.e., if} \quad 1 = \int_0^1 kx(1-x)e^x dx = k(3-e)$$

$$\therefore \quad k = \frac{1}{3-e}$$

$$\text{Mean } (\mu) = E(X) = \int_{-\infty}^{\infty} xP(x)dx$$

$$= k\int_0^1 x^2(1-x)e^x dx$$

$$= k\int_0^1 (x^2 - x^3)e^x dx$$

Integrating by parts

$$\mu = k(3e-8) = \frac{3e-8}{3-e} = 0.5496 \qquad (\because \; e = 2.718)$$

Also

$$\text{Var}(X) = \int_{-\infty}^{\infty} (x-\mu)^2 P(x)dx$$

$$= k\int_0^1 (x-\mu)^2 (x-x^2)e^x dx$$

$$= -\frac{1}{3-e}\int_0^1 \left[x^4 - x^3(1+2\mu) + x^2\mu(\mu+2) - x\mu^2\right] x\,dx$$

Integrating by parts and using $\mu = 0.5496$ we get

$$= 0.0507$$

$$\therefore \quad \text{Var}(X) = 0.0507 \quad \text{and} \quad SD = \sigma = 0.2252$$

7.3 PROBABILITY DISTRIBUTION

7.3.1 Binomial Distribution

Let X be a discrete random variable with probability function

$$P(x) = \begin{cases} {}^nC_n P^x q^{n-x} & x: 0,1,2\ldots n \\ 0 & \text{otherwise} \end{cases}$$

This probability function is called binomial probability function and the corresponding distribution is called binomial distribution if

(i) $B(n, p, x) = p(x) > 0, \quad \text{for } x = 0,1,2\ldots n$

(ii) $\sum_{x=0}^{n} B(n, p,x) = 1$

$$\text{i.e.,} \quad \sum_{x=0}^{n} B(n, p,x) = q^n + {}^nC_1 q^{n-1}p + \ldots + p_n$$

$$= (q+p)^n$$

$$= 1 \qquad \because \quad p+q=1$$

This distribution is called binomial distribution for simple reason that the probabilities are the successive terms in the expansion of $(q+p)^n$.

7.3.1.1 *Mean and Variance*

Q: Obtain the mean and variance of the binomial distribution.

Proof. Consider the mathematical expectation

$$E(X) = \sum_{x=0}^{n} xp(x)$$

$$= \sum_{x=0}^{n} x\,{}^nC_x\, p^x q^{n-x}$$

$$= \sum_{x=0}^{n} x \frac{n}{x}\, {}^{n-1}C_{x-1}\; p\, p^{x-1} q^{n-x+1-1} \qquad \left(\because\; {}^nC_x = \frac{n}{x}\,{}^{n-1}C_{x-1} = \frac{n}{x}\frac{n-1}{(x-1)}\,{}^{n-2}C_{x-2}\right)$$

$$= np \sum_{x=1}^{n} {}^{n-1}C_{x-1}\; p^{x-1} q^{n-1-(x-1)}$$

$$= np(q+p)^{n-1}$$

$$= np \qquad \because \quad p+q=1$$

$\therefore$ $\boxed{\text{Mean} = \mu = np}$

Variance:

Consider, $\text{Var}(X) = E(X^2) - [E(X)]^2$

Now, $$E(X^2) = \sum_{x=0}^{n} x^2 p(x)$$

$$= \sum_{x=0}^{n} [x(x-1)+x]p(x)$$

$$= \sum_{x=0}^{n} x(x-1)p(x) + \sum_{x=0}^{n} xp(x)$$

$$= \sum_{x=0} x(x-1)\frac{n}{x}\frac{n-1}{(x-1)}\,{}^{n-2}C_{x-2}\, p^2 p^{x-2} q^{n-2-x+2} + np$$

$$= n(n-1)p^2 \sum_{x=2}^{n} {}^{n-2}C_{x-2}\; p^{n-2} q^{n-2-(x-2)} + np$$

$$= n(n-1)p^2 (q+p)^{n-2} + np$$

$$= n(n-1)p^2 + np$$

$$= n^2p^2 - np^2 + np$$

$$\therefore \quad \text{Var}(X) = E(X^2) - [E(X)]^2$$

$$= n^2p^2 - np^2 + np - n^2p^2$$

$$= np(1-p)$$

$$= npq$$

$$\therefore \quad \sigma^2 = npq \quad \text{and} \quad SD = \sigma = \sqrt{npq}$$

Note: $N\ {}^nC_x\ q^{n-x}\ p^x$ is called binomial frequency distribution (expected values).

Example 1: The probability that a man aged 60 will live to be 70 is 0.65. What is the probability that out of 10 men, now 60 at least 7 will live to be 70?

Solution: Here $p = 0.65$ and $q = 1 - p = 0.35$ and $n = 10$

$$P(x) = {}^nC_x\ q^{n-x}\ p^x$$

Probability that at least 7 will live to be 70

$$= P(X \geq 7)$$

$$= P(7) + P(8) + P(9) + P(10)$$

$$= {}^{10}C_7(.35)^3(.65)^7 + {}^{10}C_8(.35)^2(.65)^8 + {}^{10}C_9(.35)^1(.65)^9 + {}^{10}C_{10}(.65)^{10}$$

$$= 0.514$$

Example 2: Everyone out of 15 telephone calls between 2 pm and 4 pm of a week is busy. Find the probability that out of 6 randomly selected telephone numbers are called: (i) exactly two, (ii) not more than 3, (iii) at least 3 of them are busy?

Solution: Here $p = \frac{1}{15}$, $q = 1 - p = \frac{14}{15}$ and $n = 6$

$$\therefore \quad P(x) = {}^nC_x\ q^{n-x}\ p^x$$

= Probability of x telephone calls out of n are busy

(i) Exactly two

$$P(X=2) = {}^6C_2\left(\frac{14}{15}\right)^4\left(\frac{1}{15}\right)^2 = 0.505.$$

(ii) $P(X \leq 3) = P(0) + P(1) + P(2) + P(3)$

$$= {}^6C_0\left(\frac{14}{15}\right)^6\left(\frac{1}{15}\right)^0 + {}^6C_1\left(\frac{14}{15}\right)^5\left(\frac{1}{15}\right)^1 + {}^6C_2\left(\frac{14}{15}\right)^4$$

$$\left(\frac{1}{15}\right)^2 + {}^6C_3\left(\frac{14}{15}\right)^3\left(\frac{1}{15}\right)^3$$

$$= 0.999$$

(iii) Atleast 3,

$$P(X \geq 3) = P(3) + P(4) + P(5) + P(6)$$

$$= 1 - [P(0) + P(1) + P(2)]$$

$$= 1 - \left[{}^6C_0\left(\frac{14}{15}\right)^6\left(\frac{1}{15}\right)^0 + {}^6C_1\left(\frac{14}{15}\right)^5\left(\frac{1}{15}\right)^1\right.$$

$$\left. + {}^6C_2\left(\frac{14}{15}\right)^4\left(\frac{1}{15}\right)^2\right]$$

$$= 0.005$$

Example 3: In sampling a large number of parts manufactured by a machine, the mean number of defectives in a sample of 20 is 2 out of 1000 such samples, how many would be expected to contain at least 3 defective parts.

Solution: Given mean $= 2$ and $n = 20$

$$\therefore \quad \mu = np \quad \therefore \quad 2 = 20p$$

$$\therefore \quad p = \frac{1}{10} \quad \text{and} \quad q = 1 - p = \frac{q}{10}$$

$$\therefore \quad P(x) = {}^nC_x\, q^{n-x}\, p^x,\ x = 0, 1, 2, 3 \ldots 20$$

Probability that at least 3 defectives $= P(X \geq 3)$

$$= P(3) + P(4) + P(5) + \ldots + P(20)$$

$$= 1 - [P(0) + P(1) + P(2)]$$

$$= 1 - \left[{}^{20}C_0(.9)^{20}(.1)^0 + {}^{20}C_1(.9)^{19}(.1)^1 + {}^{20}C_2(.9)^{18}(.1)^2\right]$$

$$= 0.323$$

Thus, the number of samples having at least 3 defective parts out of 1000 samples

$$= 1000 \times 0.323$$

$$= 323 \text{ samples.}$$

Example 4: The probability that a watch manufactured by a company will be defective is $\frac{1}{10}$. If 12 such watches are manufactured find the probability that:

(i) Exactly two watches will be defective
(ii) At least two watches will be defective
(iii) None will be defective

Solution: The probability of defective watch $= p = \frac{1}{10} = 0.1,\ q = 1 - p = 0.9$ and $n = 12$.

$$\therefore \quad P(x) = {}^nC_x\, q^{n-x}\, p^x \quad x = 0, 1, 2 \ldots 12$$

(i) The probability that exactly two watches will be defective

$$= P(X = 2) = {}^{12}C_2(.9)^{10}(.1)^2$$

$$= 0.2301$$

(ii) The probability that at least two will be defective $P(X \geq 2)$

$$= P(2) + P(3) + P(4) + \ldots + P(12)$$

$$= 1 - [P(0) + P(1)]$$

$$= 1 - \left[{}^{12}C_0(\ 9)^{12} + {}^{12}C_1(.9)^{11}(.1)^1\right]$$

$$= 0.3412$$

(iii) The probability that none will be defective

$$P(X = 0) = {}^{12}C_0(.9)^{12}(.1)^0$$

$$= 0.2833$$

Example 5: In a large consignment of electric lamps 5% are defective. A random sample of 8 lamps is taken for inspection. What is the probability that it has one or more defectives.

Solution: The probability of defective lamps $= 5\% = \dfrac{5}{100} = 0.05,\ q = 1 - p$ $= 0.95$ and $n = 8$.

$$\therefore \quad P(x) = {}^{n}C_{x}\, q^{n-x}\, p^{x} \quad x: 0, 1, 2 \ldots 8.$$

The probability that it has one or more defectives

$$= P(X \geq 1)$$

$$= P(1) + P(2) + \ldots + P(8)$$

$$= 1 - P(0)$$

$$= 1 - {}^{8}C_{0}(.95)^{5}$$

$$= 1 - (.95)^{5} = 0.2262$$

Example 6: Fit a binomial distribution for the following data and compare the theoretical frequencies with actual ones.

x:	5	4	3	2	1	0
f:	7	19	35	24	8	3

Solution: The data gives the observed frequencies and we need to calculate the expected frequencies.

$$\text{For mean } = \frac{\Sigma f \times x}{\Sigma f} = \frac{272}{96} = 2.9$$

$$\therefore \quad \mu = np \quad \therefore \quad 2.9 = 6p \quad \therefore \quad p \simeq 0.5 \text{ and } q = 0.5$$

The binomial distribution of fit is $N(q + p)^{n}$

$$= 96(.5 + .5)^{5}$$

The theoretical frequencies of getting 5, 4, 3, 2, 1, 0 are respectively the successive terms of the binomial expansion.

They are

$$96\left[{}^5C_0(.5)^5 + {}^5C_1(.5)^4(.5)^1 + {}^5C_2(.5)^3(.5)^2 + {}^5C_3(.5)^2(.5)^3 + {}^5C_4(.5)^4(.5)^1 + {}^5C_5(.5)^5\right]$$

$$= 3,\ 15,\ 30,\ 30,\ 15,\ 3$$

Thus, we have the table of observed and expected frequencies.

$O.f.$:	7	19	35	24	8	3
$E.f.$:	3	15	30	30	15	3

Example 7: Fit a binomial distribution to the following frequency distribution and find the expected frequencies.

x:	0	1	2	3	4	5	6	7	8	9	10
f:	6	20	28	12	8	6	0	0	0	0	0

Solution: Here $n = 10$ and $N = \sum f_i = 80$

$$\therefore \quad \text{Mean} = \mu = \frac{\Sigma f_i x_i}{\Sigma f_i} = \frac{20 + 56 + 36 + 32 + 30}{80} = 2.175$$

$$\therefore \quad np = 2.175$$

$$\therefore \quad p = \frac{2.175}{10} = 0.2175 \quad q = 1 - p = 0.7825$$

Hence, the binomial distribution to be fitted is

$$N(q+p)^{10} = 80(.7825 + .2175)^{10}$$

$$= 80\left[{}^{10}C_0\, q^{10} + {}^{10}C_1\, q^9 p + {}^{10}C_2\, q^8 p^2 + \ldots + {}^{10}C_{10} p^{10}\right]$$

$\therefore$ The expected or theoretical frequencies are

x:	0	1	2	3	4	5	6	7	8	9	10
f:	6.9	19.1	24.0	17.8	8.6	2.9	.7	.1	0	0	0

7.3.2 Poisson Distribution

Definition: A discrete random variable X is said to follow the poisson distribution if its probability function is given by

$$P(x) = \frac{e^{-\mu}\,\mu^x}{x!} : x = 0, 1, 2, 3 \ldots \infty$$

Theorem 1: Prove that the Poisson distribution is a limiting form of the binomial distribution.

Proof. The probability function of a random variable X following the binomial distribution is given by

$$B(n,p,x) = {}^nC_x \, p^x \, q^{n-x}$$

$$= \frac{n(n-1)(n-2)\ldots(n-x+1)}{x!} p^x \, q^{n-x}$$

$$= \frac{n(n-1)(n-2)\ldots(n-x+1)}{x!} p^x (1-p)^{n-x}$$

We have $\mu = np \quad \therefore \quad p = \frac{\mu}{n}$ and $q = 1 - \frac{\mu}{n}$

$$= \frac{n(n-1)(n-2)\ldots(n-x+1)}{x!} \left(\frac{\mu}{n}\right)^x \left(1 - \frac{\mu}{n}\right)^{n-x}$$

$$= \frac{\mu^x}{x!}\left[1\left(1-\frac{1}{n}\right)\left(1-\frac{2}{n}\right)\ldots\left(1-\frac{x+1}{n}\right)\right]\left(1-\frac{\mu}{n}\right)^n\left(1-\frac{\mu}{n}\right)^{-x}$$

$$= \frac{\mu^x}{x!}\left[1\left(1-\frac{1}{n}\right)\left(1-\frac{2}{n}\right)\ldots\left(1-\frac{x+1}{n}\right)\right]\left[\left(1-\frac{\mu}{n}\right)^{\frac{n}{\mu}}\right]^{\mu}\left(1-\frac{\mu}{n}\right)^{-x}$$

Now, let, $n \to \infty$ such that $np = \mu$ remains fixed and finite which is possible only when $p \to 0$ and $n \to \infty$; when $n \to \infty$, $\left(1-\frac{1}{n}\right), \left(1-\frac{2}{n}\right) \ldots \left(1-\frac{x+1}{n}\right)$ and $\left(1-\frac{\mu}{n}\right)^{-x}$ all approach to 1 and

$$\lim_{n\to\infty}\left[\left(1-\frac{\mu}{n}\right)^{\frac{n}{\mu}}\right]^{\mu} = e^{-\mu}$$

$$\therefore \quad \lim_{n\to\infty} B(n,p,x) = \frac{e^{-\mu}\,\mu^x}{x!} : x = 0, 1, 2, \ldots \infty$$

which is known as Poisson distribution.

Poisson distribution function is denoted by $P(\mu, x)$ and is defined as

$$P(\mu, x) = \frac{e^{-\mu}\,\mu^x}{x!} : x = 0, 1, 2, \ldots \infty$$

Provided

(i) $P(\mu, x) \geq 0$

(ii) $\sum_{i=1}^{\infty} P(\mu, x) = 1$

Theorem 2: Obtain the mean and variance of the Poisson distribution.

Proof. Consider

$$E(X) = \sum_{x=0}^{\infty} xp(x) = \sum_{x=0}^{\infty} \frac{x\, e^{-\mu}\, \mu^x}{x!}$$

$$= e^{-\mu} \sum_{x=1}^{\infty} \frac{\mu \cdot \mu^{x-1}}{(x-1)!}$$

$$= e^{-\mu} \mu \left(1 + \frac{\mu}{1!} + \frac{\mu^2}{2!} + \frac{\mu^3}{3!} + \dots \right)$$

$$= e^{-\mu} \cdot \mu \cdot e^{\mu}$$

$$= \mu$$

$$\therefore \quad \text{Mean} = E(X) = \mu$$

Next

$$\text{Var}(X) = E(X^2) - [E(X)]^2$$

Consider, $E(X^2) = \sum_{x=0}^{\infty} x^2 p(x)$

$$= \sum_{x=0}^{\infty} [x(x-1) + x] \frac{e^{-\mu}\, \mu^x}{x!}$$

$$= \sum_{x=0}^{\infty} x(x-1) \frac{e^{-\mu}\, \mu^{x-2}\, \mu^2}{x(x-1)(x-2)!} + \sum_{x=0}^{\infty} \frac{x\, e^{-\mu} \mu^x}{x!}$$

$$= e^{-\mu}\, \mu^2 \sum_{x=2}^{\infty} \frac{\mu^{x-2}}{(x-2)!} + \mu$$

$$= e^{-\mu} \cdot \mu^2\, e^{\mu} + \mu$$

$$= \mu^2 + \mu$$

$$\therefore \quad \text{Var}(X) = \mu^2 + \mu - (\mu)^2$$

$$= \mu$$

For Poisson distribution mean and variance are equal.

i.e., mean $= \mu$ and $\text{Var}(X) = \mu \quad \therefore \quad \text{SD} = \sigma = \sqrt{\mu}$.

Note: If $n \geq 30$ a sample is called large otherwise small. In case of large samples we use Poisson distribution and binomial distribution in case of small samples.

Example 1: If the probability of a bad reaction from certain injection is 0.001. Determine the chance that out of 2000 individuals more than two will get a bad reaction.

Solution: It follows a Poisson distribution as the probability of occurrence is very small and n is large.

$$\text{Mean} = \mu = np = 2000 \times .001 = 2$$

$$\therefore \quad P(X) = \frac{e^{-\mu}\mu^x}{x!}; \quad x = 0\ 1\ 2 \ldots 2000$$

Probability that more than 2 will get a bad reaction.

$$= P(X > 2) = P(3) + P(4) + P(5) + \ldots + P(2000)$$

$$= 1 - [P(0) + P(1) + P(2)]$$

$$= 1 - \left[e^{-2} + 2e^{-2} + \frac{2^2}{2!}e^{-2}\right]$$

$$= 1 - \frac{5}{e^2} = 0.32.$$

Example 2: In a certain factory turning out razor blades there is a small change of 0.002 for any blade to be defective. The blades are supplied in packets of 10. Use Poisson distribution to calculate the approximate number of packets containing no defective, one defective and two defective blades respectively in a consignment of 10,000 packets.

Solution: We have

$$\mu = \text{Mean} = np.$$

$$\therefore \quad \mu = 10 \times 0.002 = 0.02$$

$$\therefore \quad P(X) = \frac{e^{-\mu}\mu^X}{X!}; \quad X = 0, 1, 2, 3, \ldots$$

$$= \frac{e^{-0.02}(0.02)^X}{X!}$$

(i) Probability of no defective

$$P(X = 0) = \frac{e^{-0.02}(.02)^0}{0!} = 0.9802$$

$\therefore$ Number of packets containing no defective blades

$$= 10,000 \times 0.9802$$

$$= 9802$$

(ii) Probability of one defective blade

$$P(X = 1) = e^{-0.02}\frac{(.02)^1}{1!}$$

$$\therefore \quad \text{Number of packets} = 10,000 \times e^{-0.02}(0.02) = 196$$

(iii) Probability of two defective blades

$$P(X = 2) = \frac{e^{-0.02}(0.2)^2}{2!}$$

$$\therefore \quad \text{Number of packets} = 10,000 \times \frac{e^{-0.02}(.02)^2}{2!} = 2 \text{ approx.}$$

Example 3: The average number of telephone calls booked at an exchange between 10 AM and 10.10 AM is 4. Find the probability that on a randomly selected day 2 or more calls are booked during this time gap. On how many days of a year would you expect booking of 2 or more calls during that time gap?

Solution: Given $\mu = 4$

$$P(X) = \frac{e^{-\mu}\mu^X}{X!} = \frac{e^{-4}(4)^X}{X!}$$

$\therefore$ Probability that on a randomly selected day 2 or more calls are booked between 10 AM and 10.10 AM is

$$P(X \geq 2) = P(2) + P(3) + P(4) + \ldots$$
$$= 1 - [P(0) + P(1)]$$
$$= 1 - \left[e^{-4} + 4e^{-4}\right] = 0.9084218$$

$\therefore$ Number of days in a year we can except booking 2 or more calls

$$= 365 \times 0.9084218$$
$$= 331.517 = 331 \text{days}$$

Example 4: A manufacturer knows that 5% of items of his product is defective. If he sells the items in boxes of 100 and guarantees that not more than 4 items will be defective. What is the probability that a box will fail to meet the guaranteed quality?

Solution: Let $P = 5\% = \dfrac{5}{100} = 0.05$ and $n = 100$

Since P is small and n is large we use Poisson distribution

$$P(X) = \frac{e^{-\mu}\mu^X}{X!}$$
$$\therefore \quad \mu = np = 100 \times .05 = 5$$

Probability that guaranteed quality is not met

$$= P(X > 4) = 1 - P(X \leq 4)$$
$$= 1 - e^{-5}\left[1 + 5 + \frac{5^2}{2!} + \frac{5^3}{3!} + \frac{5^4}{4!}\right]$$
$$= 1 - 0.4405$$
$$= 0.5595$$

Example 5: An electric utility company has found that the weekly number of occurrences of lightening striking the transformer is a Poisson distribution with $\lambda = 5.8$. Find the probability that: (i) no transformer will be struck in a week and (ii) one transformer will be struck.

Solution: Here $\lambda = 5.8$ and Poisson distribution is

$$P(x) = \frac{e^{-\lambda}(\lambda)^x}{x!}$$

(i) Probability of no transformer will be struck

$$P(X=0) = \frac{e^{-5.8}(5.8)^0}{0!} = e^{-5.8} = \frac{1}{e^{5.8}} = 0.00030$$

(ii) Probability of one transformer will be struck

$$P(X=1) = \frac{e^{-5.8}(5.8)^1}{1!} = \frac{5.8}{e^{5.8}} = 0.01755.$$

Example 6: The frequency of accidents per shift in a factory is shown in the following table.

Accidents per shift:	0	1	2	3	4	Total
Freq:	192	100	24	3	1	= 300

Calculate the mean number of accidents per shifts. Find the corresponding Poisson distribution.

Solution:

$$\text{Mean} = \mu = \frac{\Sigma f(x)}{\Sigma f(x)} = \frac{161}{320} = 0.5031 \simeq 0.50$$

$$\text{Now,} \quad \sigma^2 = \frac{\Sigma x^2 f(x)}{\Sigma f(x)} - \left[\frac{\Sigma x f(x)}{\Sigma f(x)}\right]^2$$

$$= \frac{239}{320} - \left(\frac{161}{320}\right)^2 = 0.4938$$

Hence we fit Poisson distribution to the given data.

Consider $\mu = \bar{x} = 0.5$ and

$$P(\mu x) = \frac{e^{-\mu}\mu^x}{x!}$$

$$\therefore \quad P(0.5, x) = \frac{e^{-0.5}(.5)^x}{x!} \quad x: 0, 1, 2, 3, 4$$

$$\text{For } x = 0 \quad P(0) = 0.6065$$

$x = 1 \quad P(1) = 0.3033$

$x = 2 \quad P(2) = 0.0758$

$x = 3 \quad P(3) = 0.0126$

$x = 4 \quad P(4) = 0.0016$

Multiplying these by total frequency 320 we get

Accident per shift:	0	1	2	3	4
Freq:	194	97	24	4	1

which is closer to the observed data.

7.3.3 Normal Distribution

Definition: A continuous random variable X having probability density function

$$f(x) = \frac{1}{\sigma\sqrt{2\pi}} e^{\frac{-(x-\mu)^2}{2\sigma^2}}, \quad -\infty < x < \infty$$

is said to have a normal or Gaussian distribution with mean μ and standard deviation σ (or variance σ^2).

By the definition of $f(x)$ it is clear that

(i) $f(x) = N(\mu, \sigma, x) > 0$

(ii) $\int_{-\infty}^{\infty} N(\mu, \sigma, x)dx = 1$

(iii) $f(x) = \frac{1}{\sigma\sqrt{2\pi}} e^{\frac{-(x-\mu)^2}{2\sigma^2}} \quad -\infty < x < \infty$ is bell shaped.

(iv) The normal curve is symmetrical about its mean.

(v) Approximately 95% of the distribution lies within two standards of the mean.

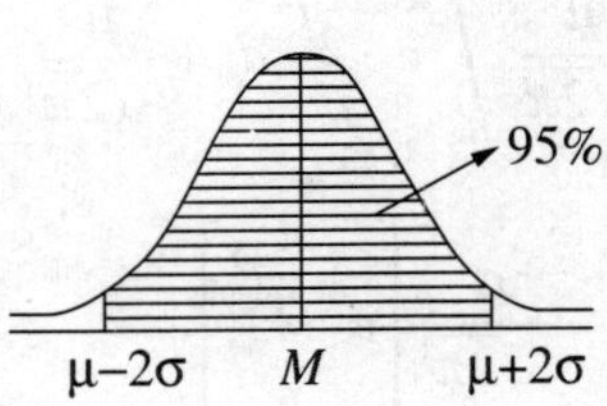

(vi) Nearly 99.8% (almost all) of the distribution lies within 3σ of the mean.

(vii) The maximum value of $f(x)$ occurs when $x = \mu$ and is given by

$$f(x) = \frac{1}{\sigma\sqrt{2\pi}}$$

(viii) The points of inflection occur at $x = \mu - \sigma$ and $x = \mu + \sigma$.

(ix) The actual size of the bell-shaped normal curve depends on the value of μ and σ.

7.3.3.1 *Mean and variance of the normal distribution*

If x is a probability density function then

$$E(x) = \int_{-\infty}^{\infty} x\, f(x) dx$$

$$= \frac{1}{\sigma\sqrt{2\pi}} \int_{-\infty}^{\infty} x e^{\frac{-(x-\mu)^2}{2\sigma^2}} dx$$

Put $t = \dfrac{x-\mu}{\sigma}$ and we assume that

$$\frac{1}{2\pi} \int_{-\infty}^{\infty} e^{-\frac{1}{2}t^2} dt = 1$$

$\therefore \quad x = t\sigma + \mu$ and $\dfrac{dx}{dt} = \sigma$

When $x \to \infty$, $t \to \infty$ and $x \to -\infty$, $t \to -\infty$

$$\therefore \quad E(x) = \frac{1}{\sigma\sqrt{2\pi}} \int_{-\infty}^{\infty} (\mu + \sigma t) e^{-\frac{1}{2}t^2} \sigma dt$$

$$= \frac{\mu}{\sqrt{2\pi}} \int_{-\infty}^{\infty} e^{-\frac{1}{2}t^2} dt + \frac{\mu}{\sqrt{2\pi}} \int_{-\infty}^{\infty} t e^{-\frac{1}{2}t^2} dt.$$

$$= \mu + \frac{\sigma}{\sqrt{2\pi}} \left[e^{-\frac{1}{2}t^2} \right]_{-\infty}^{\infty} \quad \because \quad \text{area} = 1$$

$$= \mu + 0$$

$$\therefore \quad E(x) = \text{Mean} = \mu.$$

(ii) Variance $= \sigma^2$

By definition,

$$\text{Var}(x) = \int x^2 f(x) dx - \mu^2$$

$$= I - \mu^2$$

Now,

$$I = \frac{1}{\sigma\sqrt{2\pi}} \int_{-\infty}^{\infty} (\mu + \sigma t)^2 e^{-\frac{1}{2}t^2} \sigma dt$$

$$= \frac{1}{\sqrt{2\pi}} \left[\mu^2 \int_{-\infty}^{\infty} e^{-\frac{1}{2}t^2} dt + 2\mu\sigma \int_{-\infty}^{\infty} t\, e^{-\frac{1}{2}t^2} dt + \sigma^2 \int_{-\infty}^{\infty} t^2 e^{-\frac{1}{2}t^2} dt \right]$$

Consider

$$\int_{-\infty}^{\infty} t^2 e^{-\frac{1}{2}t^2} dt = \int_{-\infty}^{\infty} t \left(te^{-\frac{1}{2}t^2} \right) dt$$

$$= \left[te^{-\frac{1}{2}t^2} \right]_{-\infty}^{\infty} - \int_{-\infty}^{\infty} -e^{-\frac{1}{2}t^2} dt$$

$$= 0 + \sqrt{2\pi}$$

$$\therefore \quad I = \frac{1}{\sqrt{2\pi}} \left[\mu^2 \sqrt{2\pi} + 2\mu\sigma \left[e^{-\frac{1}{2}t^2} \right]_{-\infty}^{\infty} + \sigma^2 \sqrt{2\pi} \right]$$

$$= \mu^2 + \sigma^2 \qquad \therefore \quad e^{-\frac{1}{2}t^2} \to 0 \quad \text{as } t \to \pm\infty$$

$$\therefore \quad \text{Var}(x) = \mu^2 \sigma^2 - \mu^2$$

$$= \sigma^2$$

$$\therefore \quad \text{Var}(x) = \sigma^2.$$

7.3.3.2 *Area under the normal curve*

By taking $Z = \frac{x - \bar{x}}{\sigma}$, the standard normal curve is formed. The total area under this curve is 1. The area under the curve is divided into two equal parts by $Z = 0$. Left hand side area and right hand side area to $Z = 0$ is 0.5. The area between the ordinates $Z = 0$ and any other ordinate can be noted from the table.

Example 1: The weekly wages of 1000 workers are normally distributed with mean of Rs 70 and SD of Rs 5. Estimate the number of workers whose weekly wages will be:

(i) Between Rs. 70 and Rs. 72
(ii) More than Rs 75 and
(iii) Less than Rs 65.

Solution: We have $Z = \frac{x - \mu}{\sigma}$. Here $\mu = 70$ and $\sigma = 5$

(i) Between x = Rs. 70 and Rs. 72

When $x = 70$, $Z = \frac{x - \mu}{\sigma} = \frac{70 - 70}{5} = 0$

When $x = 72$, $Z = \frac{x - \mu}{\sigma} = \frac{72 - 70}{5} = 0.4$

$$\therefore \quad P(70 \le X \le 72) = P(0 \le Z < 0.4)$$
$$= 0.1554.$$

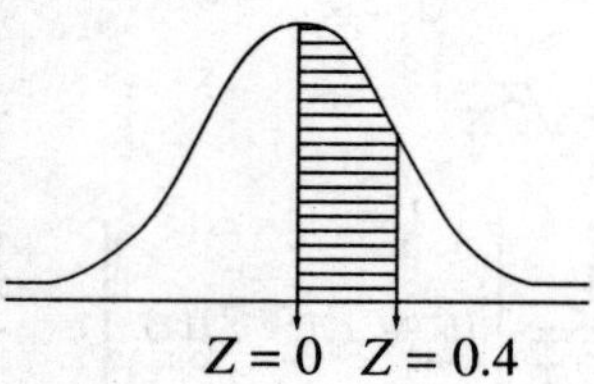

$\therefore$ Number of workers getting wages between Rs. 70 and Rs. 72

$$= 1000 \times 0.1554 = 155.4 \simeq 155.$$

(ii) When x = Rs. 75, $Z = \frac{75 - 70}{5} = 1$

$$\therefore \quad P(X > 75) = P(Z > 1)$$

$$= 0.5 - \text{Area between } Z = 0 \text{ and } Z = 1$$

$$= 0.5 - 0.3413$$

$$= 0.1587$$

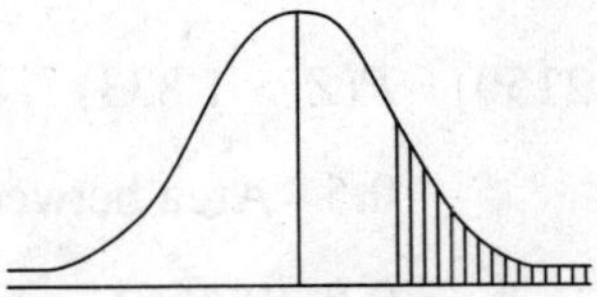

Z = 0 Z = 1

∴ Number of workers getting wages more than Rs. 75

$$= 1000 \times 0.1587 = 158.7 \simeq 159.$$

(iii) When x = Rs. 63, $Z = \dfrac{63 - 70}{5} = -1.4$

$$\therefore \quad P(X < 63) = P(Z < -1.4)$$

$$= 0.5 - \text{Area between } Z = -1.4 \text{ to } Z = 0$$

$$= 0.5 - 0.4192$$

$$= 0.0808$$

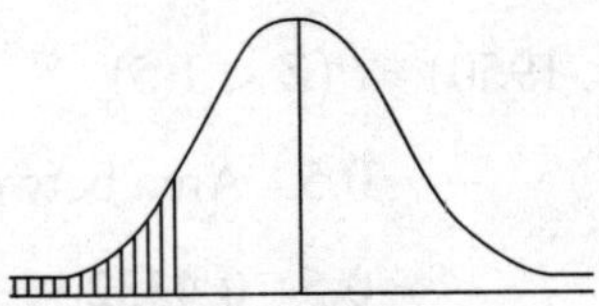

Z = −1.4 Z = 0

∴ Number of workers getting wages less than Rs. 63

$$= 1000 \times 0.0808 = 80.8 \simeq 81.$$

Example 2: In a test on 2000 electric bulbs it was found that the life of a particular make was normally distributed with an average life of 2040 hours and SD of 60 hours. Estimate the number of bulbs likely to burn for:

(i) More than 2150 hours.

(ii) Less than 1950 hours and

(iii) More than 1920 but less than 2160 hours.

Solution: Given $\mu = 2040$ hours and $\sigma = 60$ hours

(i) More than 2150 hours.

When $x = 2150$, $Z = \dfrac{x-\mu}{\sigma} = \dfrac{2150-2040}{60} = 1.833$

$$\therefore \quad P(X > 2150) = P(Z > 1.833)$$

$$= 0.5 - \text{Area between } Z = 0 \text{ and } Z = 1.833$$

$$= 0.5 - 0.4664$$

$$= 0.0336$$

$\therefore$ Number of bulbs $= 2000 \times 0.0336 = 67$ bulbs.

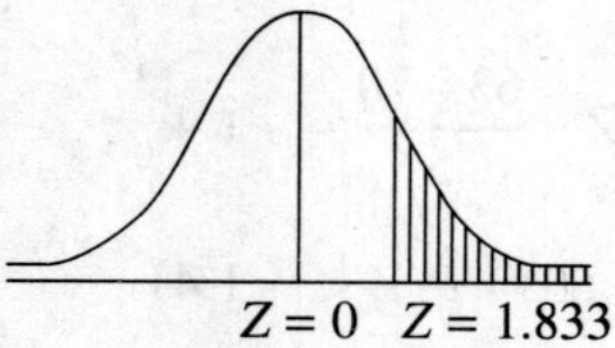

(ii) Less than 1950 hours.

When $x = 1950$, $Z = \dfrac{x-\mu}{\sigma} = \dfrac{1950-2040}{60} = -1.5$

$$\therefore \quad P(X < 1950) = P(Z < 1.5)$$

$$= 0.5 - \text{Area between } Z = -1.5 \text{ to } Z = 0$$

$$= 0.5 - 0.4332$$

$$= 0.0668$$

$\therefore$ Number of bulbs $= 0.0668 \times 2000$

$$= 133.6$$

$$= 134 \text{ bulbs}$$

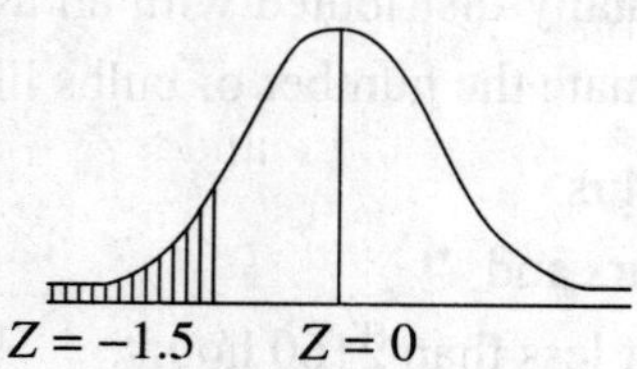

(iii) More than 1920 hours but less than 2160 hours.

When $x = 1920,\quad Z = \dfrac{1920 - 2040}{60} = 2$

When $x = 2160,\quad Z = \dfrac{2160 - 2040}{60} = -2$

$$\therefore \quad P(1920 < x < 2160)$$

$$= P(-2 < Z < +2)$$

$$= \text{Area between } Z = -2 \text{ to } Z = 0$$

$$+ \text{ area between } Z = 0 \text{ to } Z = 2$$

$$= .4772 + .4772$$

$$= 0.9544$$

$\therefore$ Number of bulbs $= 0.9544 \times 2000 = 1909$ nearly.

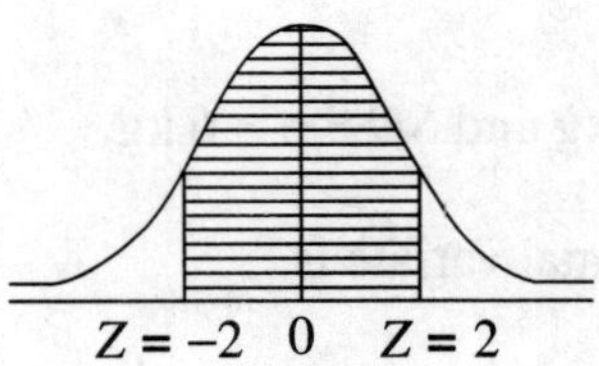

Example 3: In a normal distribution 31% of the items are under 45 and 8% are over 64. Find the mean and SD of the distribution.

Solution: Let Z_1 and Z_2 be the values of Z.

When $x = 45,\quad Z_1 = \dfrac{45 - \mu}{\sigma}$ and

When $x = 64,\quad Z_2 = \dfrac{64 - \mu}{\sigma}$

Given $\displaystyle\int_{-\infty}^{Z_1} \phi(Z)dz = 0.31$ and $\displaystyle\int_{Z_2}^{\infty} \phi(Z)dz = 0.08$

From the normal curve table

$$Z_1 = -0.5 \quad \text{and} \quad Z_2 = 1.4$$

$$\therefore \quad -0.5 = \frac{45 - \mu}{\sigma}, \quad 1.4 = \frac{64 - 14}{\sigma}$$

Solving, we get $\mu = 50$ and $\sigma = 10$.

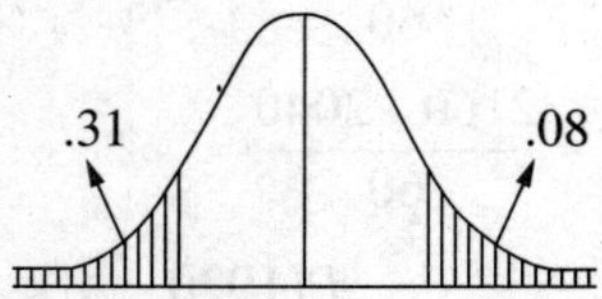

Example 4: The mean weight of 500 students in a certain college is 50 kg and the standard deviation is 6 kg. Assuming that the weights are normally distributed, find the number of students weighing: (i) between 40 and 50 kg and (ii) more than 60 kg, given that $A(1.6667) = 0.4525$ where

$$A(Z) = \frac{1}{\sqrt{2\pi}} \int_0^1 e^{-\frac{1}{2}Z^2} dZ$$

Solution: Given $\mu = 50$ kg and $SD = \sigma = 6$ kg.

$\therefore$ The standard normal variate is

$$Z = \frac{x - \mu}{\sigma} = \frac{x - 50}{6}$$

When $x = 60$, we have $Z = \dfrac{60 - 50}{6} = 1.6667$

When $x = 40$, we have $Z = \dfrac{40 - 50}{6} = -1.6667$

and when $x = 50$, we have $Z = \dfrac{50 - 50}{6} = 0$

(i) The probability that a student selected at random weighing between 40 and 50 kg is

$$P(40 \le x \le 50) = P(-1.6667 < Z < 0)$$

$$= A(1.6667) = 0.4525.$$

$\therefore$ Number of students $= 500 \times 0.4525 = 226.$

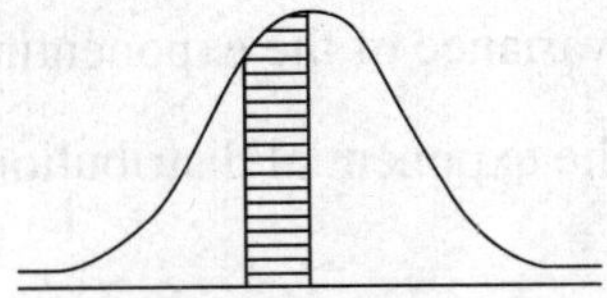

$Z = -1.6667 \quad Z = 0$

(ii) The probability that a student selected at random weighing over 60 kg is

$$
\begin{aligned}
P(x \geq 60) &= P(Z > 1.6667) \\
&= 0.5 - P(0 \leq Z \leq 1.6667) \\
&= 0.5 - 0.4525 \\
&= 0.0475
\end{aligned}
$$

$$
\begin{aligned}
\therefore \quad \text{Number of students} &= 0.0475 \times 500 \\
&= 24
\end{aligned}
$$

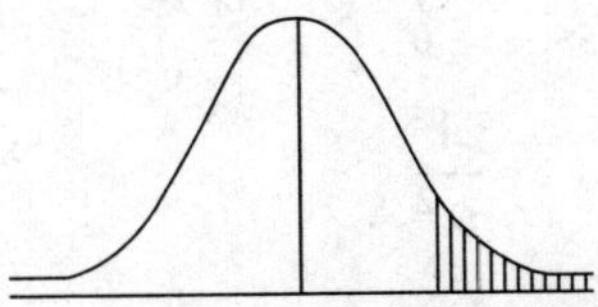

$Z = 0 \quad Z = 1.6667$

7.3.4 Exponential Distribution

Definition: A continuous random variable x assuming all non-negative values is said to have an exponential distribution with parameter $\alpha > 0$, if its probability density function denoted by $e(\alpha,x)$ is given by

$$
e(\alpha,x) = \begin{cases} \alpha e^{-\alpha x} & \text{for } x > 0 \\ 0 & \text{for } x \leq 0 \end{cases}
$$

We have

(i) $e(\alpha,x) > 0$

(ii) $\displaystyle\int_{-\infty}^{\infty} e(\alpha,x)dx = \int_{0}^{\infty} \alpha e^{-\alpha x} dx = 1$

Q: Obtain the mean and variance of the exponential distribution.

Solution: The mean of the exponential distribution is given by

$$E(x) = \int_{-\infty}^{\infty} xe(\alpha,x)dx$$

$$= \int_{0}^{\infty} \alpha xe^{-\alpha x}dx = \frac{1}{\alpha}$$

$$\therefore \quad E(x) = \frac{1}{\alpha} = \text{mean.}$$

Next

$$\text{Var}(x) = \int_{-\infty}^{\infty} (x-\mu)^2 e(\alpha,x)dx$$

$$= \alpha \int_{0}^{\infty} \left(x - \frac{1}{\alpha}\right)^2 e^{-\alpha x}dx$$

$$= \frac{1}{\alpha^2}$$

Hence, for exponential distribution mean = Standard deviation = $\frac{1}{\alpha}$.

Further, from the definition of $e(\alpha,x)$ we have

(i) $P(-\infty < x < \infty) = P(0 < x < \infty) = 1$

(ii) $P(0 < x < a) = \int_{0}^{a} e(\alpha x)dx$

(iii) $P(x \geq a) = P(0 \leq x < \infty) - P(0 \leq x \leq a)$

$$= 1 - \int_{0}^{a} e(\alpha,x)dx.$$

Example 1: The life time of a certain kind of battery is a random variable which as an exponential distribution with a mean of 200 hours. Find the probability that such a battery will last: (i) at most 100 hours and (ii) last anywhere from 400 to 600 hours.

Solution: Given mean $= \frac{1}{\alpha} = 200$ and

$$e(\alpha, x) = \begin{cases} \alpha\, e^{-\alpha x} & \text{for } x > 0 \\ 0 & \text{otherwise} \end{cases}$$

(i) Probability that a battery will last at most 100 hours

$$= P(x \le 100)$$

$$= \int_0^{100} \alpha\, e^{-\alpha x} dx = \left[\frac{\alpha\, e^{-\alpha x}}{-\alpha}\right]_0^{100}$$

$$= e^{-0} - e^{-100\alpha} = 1 - \frac{1}{e^{1/2}} = 0.3935$$

(ii) Probability that a battery will last between 400 and 600 hours

$$= P(400 \le x \le 600)$$

$$= \int_{400}^{600} \alpha\, e^{-\alpha x} dx = \left[\frac{\alpha\, e^{-\alpha x}}{-\alpha}\right]_{400}^{600}$$

$$= e^{-400\alpha} - e^{-600\alpha} = e^{-2} - e^{-3} = \frac{1}{e^2} - \frac{1}{e^3} = 0.085 \quad \left(\because \alpha = \frac{1}{200}\right)$$

Example 2: The sales per day in a shop is exponentially distributed with mean value of Rs. 100. If sales tax is levied at the rate of 8%, what is the probability that the sales tax return from that shop will exceed Rs. 30 on two consecutive days.

Solution: Given mean value of sales tax = Rs. 8

i.e., $\frac{1}{\alpha} = 8 \quad \therefore \quad \alpha = \frac{1}{8} = 0.125.$

Hence the density function

$$e(\alpha, x) = \begin{cases} 0.125 e^{-0.125x} & \text{for } x > 0 \\ 0 & \text{otherwise} \end{cases}$$

Probability that the sales tax return will exceed Rs. 30 on a day $= P(X > 30)$

$$= \int_{30}^{\infty} e(\alpha\, x) dx$$

$$= 0.125 \int_{30}^{\infty} e^{\frac{-x}{8}} dx = e^{-3.75} = P \text{ say}$$

Hence, the probability that the sales tax returns will exceed Rs. 30 on two consecutive days is

$$P^2 = (e^{-3.75}) = e^{-7.5}$$

Example 3: After the establishment of a new local T.V. station the increase in the sales per day in a T.V. showroom is exponentially distributed with mean 2. If two days are selected at random what is the probability that:

(i) on both days the increase is over 10 units

(ii) the increase is over 10 units on at least one of the two days

Solution: Given mean $= \frac{1}{\alpha} = 2$

$\therefore$ The density function is

$$e(\alpha, x) = \begin{cases} \frac{1}{2} e^{\frac{-x}{2}} & \text{for } x > 0 \\ 0 & \text{otherwise} \end{cases}$$

The probability of an increase in sales of over 10 units on any one randomly selected day is

$$P(X \geq 10) = \int_{10}^{\infty} \frac{1}{2} e^{\frac{-x}{2}} dx$$

$$= e^{-5} = P \text{ say}$$

(i) Probability that the increase is over 10 units on both days is

$$= P^2 = (e^{-5})^2 = e^{-10}$$

(ii) The probability that the increase is over 10 units on at least one of the two days is

$$= {}^2C_1 pq + {}^2C_2 p^2$$
$$= 2p(1-p) + p^2$$
$$= 2p - p^2$$
$$= 2e^{-5} - e^{-10}$$

Example 4: Find the c.d.f. for an exponential distribution.

Solution: For exponential distribution the p.d.f. is given by

$$f(x) = \begin{cases} \alpha\, e^{-\alpha x} & \text{for } x > 0 \\ 0 & \text{otherwise} \end{cases}$$

c.d.f. is given by

$$F(x) = \int_{-\infty}^{x} f(x)dx$$

Since $f(x) = 0$ for $x \le 0$ we have

$F(x) = 0$ for $x \le 0$ and for $x > 0$

$$F(x) = \int_{-\infty}^{0} f(x)dx + \int_{0}^{x} f(x)dx$$

$$= 0 + \int_{0}^{x} \alpha\, e^{-\alpha x} dx$$

$$= \alpha \left[\frac{e^{-\alpha x}}{-\alpha} \right]_0^x = \frac{e^{-\alpha x} - e^{-0}}{-1} = 1 - e^{-\alpha x}$$

Hence

$$F(x) = \begin{cases} 1 - e^{-\alpha x} & \text{for } x > 0 \\ 0 & \text{otherwise} \end{cases}$$

Example 5: The length of a telephone conversation is an exponential variate with mean 3 minutes. Find the probability that a cell: (i) ends in less than 3 minutes, (ii) takes between 3 to 5 minutes.

QUESTION BANK

A. Random Variables

1. A random variable x has the following probability function.

x:	-2	-1	0	1	2	3
$P(x)$:	0.1	k	0.2	$2k$	0.3	k

Find the value of k and calculate the mean and variance.

Ans: $k = .1$ mean $= .8$, $\sigma^2 = 2.232$

2. The diameter of an electric cable is assumed to be a continuous variate with PDF

$$f(x) = 6x(1-x) \quad 0 \le x \le 1$$

Verify that the above function is a PDF. Also find the mean and variance.

Ans: $f(x)$ is p.d.f. $\bar{x} = \frac{1}{2}$, $\sigma^2 = \frac{1}{20}$

3. A function is defined as follows

$$\begin{aligned} f(x) &= 0 & x < 2 \\ &= \frac{1}{18}(2x+3) & 2 \le x \le 4 \\ &= 0 & x > 4 \end{aligned}$$

Show that it is a density function. Find the probability that a variate having this density will fall in the interval $2 \le x \le 3$.

Ans: 4/9.

4. Obtain the distribution function of the total number of heads occurring in three tosses of an unbiased coin.

Ans: $f(x) = 0 \; -\infty < x < 0$, $f(x) = \frac{1}{8} \; 0 \le x \le 1$, $f(x) = \frac{1}{2} \; 1 \le x \le 2$, $f(x) = \frac{7}{8} \; 2 \le x \le 3$, $f(x) = 1$, $3 \le x \le \infty$

5. Find the mean and variance for the following discrete distribution.

x:	8	12	16	20	24
$P(x)$:	1/8	1/6	3/8	1/4	1/12

Ans: SD $= 2\sqrt{5}$

6. The probability density $P(x)$ of a continuous random variable is given by

$$P(x) = Y_0 e^{-|x|} \quad -\infty < x < \infty$$

Prove that $Y_0 = 1/2$. Find the mean and variance of the function.

Ans: 0.2

B. Binomial Distribution and Poisson Distribution

7. The probability that a bomb dropped from a plane will strike the target is $\frac{1}{5}$. If six bombs are dropped find the probability that

   (i) exactly two will strike the target.

   (ii) at least two will strike the target.

   Ans: (i) 0.246, (ii) 0.345

8. If the probability that a newly born child is a male is 0.6 find the probability that in a family of 5 children there are exactly 3 boys.

   Ans: 0.3456.

9. The incidence of occupational disease in an industry is such that the workmen have a 20% chance of suffering from it. What is the probability that out of six workmen, 4 or more will contact the disease?

   Ans: 0.0169.

10. If on an average 8 ships out of 10 arrive safety at a port, find the mean and standard deviation of the number of ships arriving safely out of a total of 1600 ships.

    Note: $p = .8,\ q = .2$

    Ans: mean $= np = 1280 \quad \sigma = \sqrt{npq} = 16$

11. A certain screw making machine produces on an average of 2 defective screws out of 100 and packs them in boxes of 500. Find the probability that a box contains 15 defective screws.

    Ans: 0.035.

12. (a) Fit a Poisson distribution to the following data

x:	0	1	2	3	4
f:	46	38	22	9	1

Ans: Theoretical frequencies are 44, 43, 21, 7, 1

(b) Alpha-particles are emitted by a radioactive source at an average life of 5 emission in 20 minutes. What is the probability that there will be (i) exactly two emissions, (ii) at least two emissions in 20 minutes?

13. Students of a class were given a Mathematical aptitude test. Their marks were found to be normally distributed with mean 60 and standard deviation. What percent of students scored:

    (i) more than 60 marks

    (ii) less than 56 marks

    (iii) between 45 and 65 marks.

    Ans: (i) 50%, (ii) 21.2%, (iii) 84%

14. The mean height of 500 male students in a certain college is 151 cm and the standard deviation is 15 cm. Assuming the heights are normally distributed, find how many students have heights between 120 cm and 155 cm?

    Ans: 300 students.

15. In a normal distribution, 7% of the items are under 35 and 89% are under 63. What are the mean and the standard deviation of the distribution?

    Ans: $\bar{x} = 50.3$, $\sigma = 10.33$

16. In an examination taken by 500 candidates, the average and the standard deviation of marks obtained (normally distributed) are 40% and 10% Find approximately:

    (i) how many will pass, if 50% is fixed as a maximum?

    (ii) what should be the minimum if 350 candidates are to pass?

    (iii) how many have scored marks above 60%?

    Ans: (i) 79, (ii) 35%, (iii) 11

17. Assuming that the diameters of 1000 brass plugs taken consecutively from a machine form a normal distribution with mean 0.7515 cm and SD 0.0020 cm, how many of the plugs are likely to be rejected if the approved diameter is 0.752 ± 0.004 cm?

    Ans: 52

18. It is given that the age of thermostats of a particular make follow the normal law with mean 5 years and SD 2 years. 1000 units are sold out every month. How many of them will have to be replaced at the end of the second year.

    Ans: 67

19. For a normally distributed variate with mean 1 and SD 3 find the probabilities that

    (i) $3.43 \leq x \leq 6.19$ (ii) $-1.43 \leq x \leq 6.19$

    Ans: (i) 0.1644 (ii) 0.7686

20. A sample of 100 dry battery cells tested to find the length of life produced by a company and the following results are recorded: mean life = 12 hours, standard deviation = 3 hours. Assuming the data to be normally distributed, find the expected life of a dry cell.

    (i) have more than 15 hours.

    (ii) between 10 and 14 hours.

UNIT 8

Sampling Theory

8.1 SAMPLING

In our day-to-day life it becomes quite often necessary to draw some valid and reasonable conclusions concerning a large mass of individuals or objects. It becomes practically impossible to examine every individual or the entire group. Therefore we may prefer to examine a small part of this population known as *sample* with the motive of drawing some conclusion about the entire population based on the information revealed by the sample. The process of drawing a sample is called *sampling*. It is essential that a sample must be a random selection so that each member of the population has the same chance of being included in the sample. The fundamental assumption underlying the theory of sampling is *random sampling*.

A special case of random sampling in which each event has the same probability, P, of success and the chance of success of different events are independent whether previous trials have been made or not, is known as *simple sampling*.

8.1.1 Parameters and Statistics

The statistical constants of the population such as mean (μ), standard deviation (σ), etc., are called the *parameters*, and the constants for the sample drawn from the given population, i.e., mean $(\bar{x})$, standard deviation (S), etc., are called the *statistic*. In other words the parameters are the functions of the population values while statistic is the functions of sample values.

8.1.2 Sampling Distribution

Consider all possible samples of size 'n' which can be drawn from a given population at random. For each sample we compute a statistic (mean). The means of the samples will not be identical and are liable to vary. If we group these different means according to their frequencies, the frequency distribution so formed is known as sampling distribution of mean. Similarly, we can have sampling distribution of the medians, standard deviations, etc.

8.1.3 Objectives of Sampling

Sampling aims at gathering the maximum information about the population with the minimum effort, cost and time. The object of sampling studies is to obtain the best possible values of the parameters under specific conditions. Sampling determines the reliability of these estimates. The logic of the sampling theory is the logic of induction in which we pass from a particular (sample) to general (population) such a generalisation from sample to population is called *statistical inference*.

8.1.4 Standard Error

The standard deviation of the sampling distribution is called the standard error (SE). Thus the standard error of the sampling distribution of means is called the standard error of means. The standard error is used to assess the difference between the expected and observed values. The reciprocal of the standard error is called precision.

If $n > 30$, a sample is called large otherwise it is small. The sampling distribution of large samples is assumed to be normal.

8.1.5 Testing of Hypothesis

To reach decisions about the populations on the basis of sample information, we make certain assumptions about the populations involved. Such assumptions which may or may not be true, are called statistical hypothesis. By testing a hypothesis is meant a process for deciding whether to accept or reject the hypothesis. The method consists in assuming the hypothesis as correct and then computing the probability of getting the observed sample. If this probability is less than a certain preassigned value the hypothesis is rejected.

8.1.6 Procedure for Testing Hypothesis

(i) *Set up a hypothesis:* The first step in hypothesis testing is to set up a hypothesis about a population parameter. The two hypothesis in a statistical test are normally referred to as a null hypothesis and a alternative hypothesis. H_0 represents null hypothesis and H_1 represents the alternative hypothesis.

(ii) *Set up a suitable significance level:* Having set up the hypothesis, the next step is to test the validity of H_0 against that of H_1, at a certain level of significance. The confidence with which an experiment rejects or retains a null hypothesis, depends upon the significance level adopted. The significance level is expressed as a percentage 5% and 1% level of significance.

(iii) *Setting a test criterion:* The third step in hypothesis testing procedure is to construct a test criterion.

(iv) Doing computations.

(v) *Making decisions:* Finally we may draw statistical conclusions and take decisions. A statistical decision is a decision either to reject or to accept the null hypothesis.

8.1.7 Errors in Hypothesis Testing

There are some possibilities that though the sample statistic is a true representation of parameter by some unforseen factors the statistic lies in the rejection region (H_1 region) which leads to rejection of H_0. This type of error is called Type-I error and the probability of getting this error is α.

Similarly, there is a possibility that by some chance factor though H_0 is false the statistic lies in H_0 region, i.e., in (a,b) which leads to accepting H_0. This type of error is called Type-II error. The probability of getting this error is β.

When a statistical hypothesis is tested there are four possibilities.

(1) The hypothesis is true but our test rejects it (Type-I error)

(2) The hypothesis is false but our test accepts it (Type-II error)

(3) The hypothesis is true and our test accepts it (Correct decision)

(4) The hypothesis is false and our test rejects it (Correct decisions)

8.1.8 Null Hypothesis

The hypothesis formulated for the sake of rejecting it, under the assumption that it is true, is called the null hypothesis and is denoted by H_0.

To test whether one procedure is better than another, we assume that there is no difference between the procedures. Similarly to test whether there is a relationship between two variates, we take H_0 that there is no relationship.

By accepting a null hypothesis, we mean that on the basis of the statistic calculated from the sample, we do not reject the hypothesis.

In general, the positive ascertain that "There is no significant difference between the sample statistic and parameter" is called null hypothesis "H_0". The alternative hypothesis contradictory to H_0 and is denoted by H_1. If we are estimating the population mean μ from a sample mean $\bar{X}$ then

$H_0 : \mu = \bar{X}$ (Sample mean and population mean are equal)

$H_1 : \mu \neq \bar{X}$ (Sample mean and population mean are not equal)

8.1.9 Level of Significance

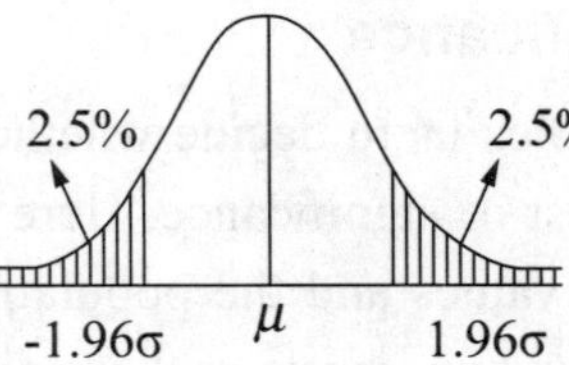

The probability level below which we reject the hypothesis is known as the level of significance.

The region in which a sample value falling is rejected, is known as the *critical region*. Generally we take two critical regions which cover 5% and 1% areas of the normal curve. The shaded part in the figure corresponds to 5% level of significance. The probability of the value of the variate falling in the critical region is the level of significance.

8.1.10 Two Tailed Test

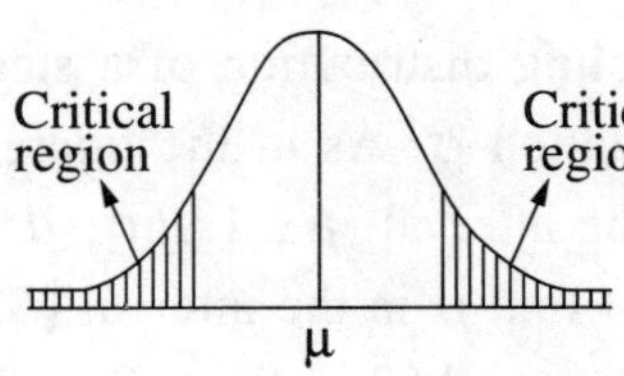

For symmetric distributions like t, Z, we generally divide the critical region (α) into two equal parts $\frac{\alpha}{2}$ in left and $\frac{\alpha}{2}$ in the right tail so that the acceptance region $(1 - \alpha)$ lies between a and b. This type of testing is called two tailed test.

8.1.11 One Tailed Test

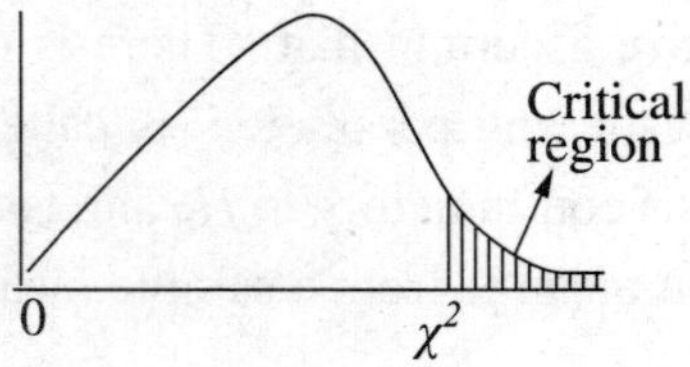

For non-symmetric distributions like χ^2, F, the critical region, is taken generally in the right tail so that the acceptance region lies between 0 and χ^2 or 0 and F_α as the case may be. This type of test is called one tailed test.

8.1.12 Test of Significance

The procedure which enables us to decide whether to accept or to reject the hypothesis is called the test of significance. Here we test whether the difference between the sample values and the population values are so large that signify evidence against the hypothesis or these differences are so small as to account for fluctuations of sampling.

8.1.13 Confidence Limits

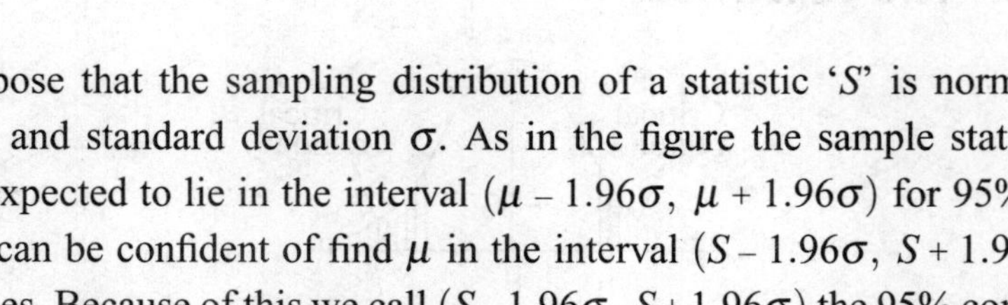

Suppose that the sampling distribution of a statistic 'S' is normal with mean μ and standard deviation σ. As in the figure the sample statistic 'S' can be expected to lie in the interval $(\mu - 1.96\sigma,\ \mu + 1.96\sigma)$ for 95% times, i.e., we can be confident of find μ in the interval $(S - 1.96\sigma,\ S + 1.96\sigma)$ the 95% coses. Because of this we call $(S - 1.96\sigma,\ S + 1.96\sigma)$ the 95% confidence interval for estimation of μ. The ends of this interval, i.e., $S \pm 1.96\sigma$ are called 95% confidence limits (or fiducial limits) for S.

Similarly, $S \pm 2.58\sigma$ are 99% confidence limits. The numbers 1.96, 2.58. etc., are called confidence coefficients and can be found from the normal curve table.

8.1.14 Degrees of Freedom

Number of degrees of freedom is the number of values in a set which may be assigned arbitrarily. For example, if $x_1 + x_2 + x_3 = 25$ and we can assign any values to two of the variables say x_1, x_2 then the value of x_3 will be known.

Therefore, the two variables are free and independent choices for finding the third. Hence, these are the degrees of freedom. If there are n observations, the degrees of freedom $(d.f)$ are $(n - 1)$. In other words, while finding the mean of a small sample, one degree of freedom is used up and $(n - 1)$ d.f are left to estimate the population variance.

8.2 STUDENT'S *t*-DISTRIBUTION

Consider a small sample of size n $(n < 30)$, drawn from a normal population with mean μ and SD σ. If $\bar{x}$ and σs be the sample mean and SD then the statistic 't' is defined as

$$t = \frac{\bar{x} - \mu}{\left(\frac{\sigma}{\sqrt{n}}\right)} \quad \text{or} \quad t = \frac{\bar{x} - \mu}{\left(\frac{\sigma s}{\sqrt{n-1}}\right)}$$

where $v = n - 1$ denotes the d.f of t.

If we calculate t for each sample, we obtain the sampling distribution for 't'. This distribution known as student's 't' distribution is given by

$$y = \frac{y_0}{\left(1 + \frac{t^2}{v}\right)} \left(\frac{v+1}{2}\right) \tag{1}$$

where y_0 is constant such that the area under the curve is unity.

8.2.1 Significance Test of a Sample Mean

Given a random small sample $x_1, x_2 \ldots x_n$ for a normal population, we have to test the hypothesis that mean of the population is μ. For this we calculate:

$$t = \frac{\bar{x} - \mu}{\sigma s / \sqrt{n}} \quad \text{where} \quad \bar{x} = \frac{1}{n}\sum_{1}^{n} x_i, \quad \sigma s^2 = \frac{1}{n-1}\sum_{1}^{n}(x_i - \bar{x})^2$$

Then find the value of t for the mean to a given d.f from the table.

If the calculated value of $t > t_{0.05}$, the difference between $\bar{x}$ and μ is said to be significant at 5% level of significance.

If $t > t_{0.01}$ the difference is said to be significant at 1% level of significance.

If $t < t_{0.05}$, the data is said to be consistent with the hypothesis that μ is the mean of the population.

Example 1: Ten individuals are chosen at random from a population and their heights in inches are found to be 63, 63, 66, 67, 68, 69, 70, 70, 71, 71. Test the hypothesis that the mean height of the universe is 66″.

(Given value of $t_{0.05} = 2.262$ for 9 d.f)

Solution: Given, mean of the universe $\mu = 66''$, and $n = 10$.

We have to calculate sample mean $(\bar{x})$ and SD σs.

$$\text{Now, mean } = \frac{1}{n}\sum x_i = \frac{1}{10}[63 + 63 + 66 + 67 + 68 + 69 + 70 + 70 + 71 + 71]$$

$$\therefore \quad \bar{x} = \frac{1}{10}(678) = 67.8$$

$$\text{And} \quad \sigma s^2 = \frac{1}{n-1}\sum(x_i - \bar{x})^2$$

$$= \frac{1}{9}\left[(63 - 67.8)^2 + (63 - 67.8)^2 + \ldots + (71 - 67.8)^2\right]$$

$$= 9.067$$

$$\therefore \quad \text{SD} = \sigma = 3.011$$

$$\text{Consider,} \quad t = \frac{\bar{x} - \mu}{\sigma s/\sqrt{n}} = \frac{67.8 - 66}{(3.011)/\sqrt{10}} = 1.89$$

$$\therefore \quad t = 1.89 < 2.262 \quad \text{table value at } t_{0.05} \text{ for 9 d.f.}$$

$\therefore$ The hypothesis is accepted at 5% level of significance that the mean height of the universe is 66″.

Example 2: Nine items of a sample have the following values 45, 47, 50, 52, 48, 47, 49, 53, 51. Does the mean of these differ significantly from the assumed mean of 47.5?

Solution: Given, population mean $\mu = 47.5$ and $n = 9$. We have to calculate sample mean $(\bar{x})$ and SD σs.

x	$d = x - A = (x - 48)$	$d^2 = (x - A)^2$
45	−3	9
47	−1	1
50	2	4
52	4	16
48	0	0
47	−1	1
49	1	1
53	5	25
51	3	9
Total	10	66

$$\therefore \quad \text{Mean} = \bar{x} = A + \frac{\sum d}{n}$$

$$= 48 + \frac{10}{9} = 49.1$$

and

$$\sigma s^2 = \frac{\sum d^2}{n} - \left(\frac{\sum d}{n}\right)^2$$

$$= \frac{66}{9} - \left(\frac{10}{9}\right)^2 = \frac{494}{81}$$

$$\therefore \quad \sigma s = 2.47$$

Consider

$$t = \frac{\bar{x} - \mu}{\sigma s / \sqrt{n-1}}$$

$$= \frac{49.5 - 47.5}{2.47/\sqrt{8}} = 1.83$$

For $v = n - 1 = 9 - 1 = 8$ d.f. table value $t_{0.05} = 2.31$.

As calculated value of $t < t_{0.05}$, the value of t is not significant at 5% level of significance which implies that there is no significant difference between $\bar{x}$ and μ. Thus the test provides no evidence against the population mean being 47.5.

Example 3: A mechanist is making engine parts with the axle diameter of 0.7 cm. A random sample of 10 parts shows mean diameter 0.742 cm with a standard deviation of 0.04 cm. On the basis of this sample would you say that the work is inferior?

Solution: Set the hypothesis that the product is not inferior, i.e., there is no significant difference between $\bar{x}$ and μ.

Given, $\mu = 0.7,\ \bar{x} = 0.742,\ \sigma x = 0.040,\ n = 10.$

$$\therefore \quad t = \frac{\bar{x} - \mu}{\sigma x/\sqrt{n-1}}$$

$$= \frac{.742 - .700}{0.040/\sqrt{10-1}} = 3.16$$

For, $v = n - 1 = 10 - 1 = 9$ d.f. the table value $t_{0.05} = 2.262$. As the calculated value of $t > t_{0.05}$, the value of t is significant at 5% level of significance. This implies that $\bar{x}$ differs significantly from μ and the hypothesis is rejected. Hence, the work is inferior.

Example 4: A machine is expected to produce nails of length 3 cm. A random sample of 25 nails gave an average length of 3.1 cm with standard deviation 0.3 cm. Can it be said that the machine is producing nails as per specification.

(Given, value of student $t_{0.05}$ for 24 d.f. is 2.064)

Solution: Set the hypothesis that the machine is producing nails as per specifications.

Given data is $\mu = 3$ cm $\quad \bar{x} = 3.1$ cm $\quad \sigma s = 0.3$ cm and $n = 10$

$$\therefore \quad t = \frac{\bar{x} - \mu}{\sigma x/\sqrt{n-1}} = \frac{3.1 - 3}{0.3/\sqrt{25-1}} = 1.63$$

For $v = n - 1 = 24$ d.f. the table value $t_{0.05} = 2.064$.

The calculated value of $t < t_{0.05} = 2.064$.

$\therefore$ The hypothesis that the machine is producing nails as per specifications is accepted at 5% level of significance.

Example 5: Show that 95% confidence limits for the mean μ of the population are $\bar{x} \pm \frac{\sigma s}{\sqrt{n}} t_{0.05}$.

Deduce that for a random sample of 16 values with mean 41.5 cm and the sum of the squares of the deviation from the mean 135 cm^2 and drawn from a normal population, 95% confidence limits for the mean of the population are 39.9 and 43.1 cm.

Solution: The probability p that $t \le t_{0.05}$ is 0.95. Hence the 95% confidence limits for μ are given by

$$\left| \frac{\bar{x} - \mu}{\sigma s / \sqrt{n}} \right| \le t_{0.05}$$

$$\therefore \quad |\bar{x} - \mu| \le \frac{\sigma s}{\sqrt{n}} t_{0.05}$$

$$\therefore \quad \bar{x} - \frac{\sigma s}{\sqrt{n}} t_{0.05} \le \mu \le \bar{x} + \frac{\sigma s}{\sqrt{n}} t_{0.05}$$

Therefore, we can say with a confidence coefficient 0.95 that the confidence interval $\bar{x} \pm \frac{\sigma s}{\sqrt{n}} t_{0.05}$ contains the population mean μ.

Now, $n = 16$ and $v = n - 1 = 15$ d.f.

$$\therefore \quad \sigma s = \sqrt{\frac{135}{15}} = 3$$

Also from the table $t_{0.05} = 2.13$ for 15 d.f.

$$\therefore \quad \frac{\sigma s}{\sqrt{n}} t_{0.05} = \frac{3}{4} \times 2.13 = 1.6 \text{ approx.}$$

Hence, the required confidence limits are 41.5 ± 1.6.

i.e., $41.5 - 1.6 = 39.9$ and $41.5 + 1.6 = 43.1$ cm.

Example 6: A sample of 10 measurements of the diameter of a sphere gave a mean of 12 cm and SD 0.15 cm. Find the 95% confidence limits for the actual diameter.

Solution: Given data is $n = 10$, $\bar{x} = 12$ cm, $\sigma s = 0.15$ and also $t_{0.05} = 2.262$ for 9 d.f.

Confidence limits for the actual diameter are

$$\bar{x} \pm \frac{\sigma s}{\sqrt{n}} t_{0.05} = 12 \pm \frac{0.15}{\sqrt{10}}(2.262)$$

$$= 12 \pm 0.1073$$

$$= 11.89 \text{ to } 12.107 \text{ cm}$$

Example 7: Sample of 12 measurements of the diameter of a metal ball gave the mean $\bar{x} = 7.38$ mm with SD $\sigma s = 1.24$ mm. Find: (a) 95% and (b) 99% confidence limits for the actual diameter.

Solution: Given data is $n = 12$, $\bar{x} = 7.38$ mm, SD $= \sigma s = 1.24$ mm.

The confidence limits for the mean are

$$\bar{x} \pm \frac{\sigma s}{\sqrt{n-1}} t_c$$

where $t_c = t_{0.05}$ for 95% confidence limits and

$t_c = t_{0.01}$ for 99% confidence limits.

For $v = 11$, $t_{0.05} = 2.20$ and $t_{0.01} = 3.11$

(a) 95% confidence limits are

$$\bar{x} \pm \frac{\sigma s}{\sqrt{n-1}} t_{0.05} = 7.38 \pm \frac{1.24}{\sqrt{11}} \times 2.20$$

$$= 7.38 \pm 0.8225$$

(b) 99% confidence limits are

$$\bar{x} \pm \frac{\sigma s}{\sqrt{n-1}} t_{0.01} = 7.38 \pm \frac{1.24}{\sqrt{11}} \times 3.11$$

$$= 7.38 \pm 1.1627.$$

Example 8: Ten students are selected at random from a college and their heights in inches are found to be 31, 50, 52, 53, 52, 55, 56, 57, 58, 59. In the length of these data and the help of the 't' test decide about the acceptance or rejection of the hypothesis that the mean height of the students of the college is 54 inches at 5% level of significance.

Example 9: A sample of 10 measurements of the diameter of a sphere gave a mean $\bar{x} = 4.38$ inches and SD $= 0.06$ inches. Find: (a) 95% and (b) 99% confidence limits for the actual diameter.

8.2.2 Test of Significance for Large Samples

We know that normal distribution is the limiting form of the binomial distribution when n is large and neither p nor q is small. Suppose we take N samples each having 'n' members. Let p be the probability of success of each member and q be the failure so that $p + q = 1$. The frequencies of samples with successes 0, 1, 2, 3...n are the terms of the binomial expansion of $N(q + p)^n$. Thus, the binomial distribution is regarded as the sampling distribution of the number of successes in the sample.

Mean for the binomial distribution is np and standard deviation is $\sqrt{npq}$.

Consider the proportion of successes:

(i) Mean proportion of success $= \dfrac{np}{n} = p$.

(ii) SD or (SE) proportion of successes $= \dfrac{\sqrt{npq}}{n} = \sqrt{\dfrac{pq}{n}}$.

Let 'x' be the observed number of successes in a sample of size 'n' and $\mu = np$ be the expected number of successes. Let the associated standard normal variate Z be defined by

$$Z = \frac{x - \mu}{\sigma} = \frac{x - np}{\sqrt{npq}}.$$

(i) If $|Z| > 1.96$ the difference is significant at 5% level of significance and we reject the hypothesis.

(ii) If $|Z| > 2.58$, we conclude that the difference is highly significant at 1% level of significance and reject the hypothesis. Since p is the probability of success and $\sqrt{\dfrac{pq}{n}}$ is the standard error (SE) proportion of success then $p \pm 2.58\sqrt{\dfrac{pq}{n}}$ are the probable limits.

Example 1: A coin is tossed 1000 times and it turns up head 540 times. Decide on the hypothesis that the coin is unbiased.

Solution: Set the hypothesis that the coin is unbiased.

Probability of getting a head in one toss $= p = \dfrac{1}{2}$

Since $p + q = 1 \quad \therefore \quad q = 1 - p = \dfrac{1}{2}$.

Actual number of heads $= 540$

and expected number of heads $= np = 1000 \times \dfrac{1}{2} = 500$.

$$\text{Consider,} \quad Z = \frac{x - np}{\sqrt{npq}} = \frac{540 - 500}{\sqrt{1000 \times \frac{1}{2} \times \frac{1}{2}}} = 2.53$$

Since, calculated $Z_c = 2.53 < 2.58$, we accept the hypothesis that the coin is unbiased.

Example 2: A sample of 100 days was taken in a coastal town and it was found that on 10 days the temperature was more than 38°C, amounting to very hot temperature. Obtain the probable limits of the percentage of very hot temperature.

Solution: Probability of very hot temperature $= p = \dfrac{10}{100} = 0.1$

Since $p + q = 1, \quad q = 1 - p = 0.9$.

$$\therefore \quad \text{Probable limits} = p \pm 2.58\sqrt{\frac{pq}{n}}$$

$$= 0.1 \pm 2.85\sqrt{\frac{.1 \times .9}{100}}$$

$$= .1 \pm 0.0774$$

$$= 0.0226 \text{ and } 0.1774$$

$\therefore$ Probable limits percentage

$$= 2.26\% \text{ to } 17.74\%$$

Example 3: Result extracts revealed that in a certain college over a period of five years 725 students had passed and 615 students had failed. Test the hypothesis that success and failure are in equal proportions.

Solution:

$$\text{Total number of students} = 725 + 615 = 1340$$

$$\text{Expected number of successes} = \frac{1340}{2} = 670$$

$$\text{Observed number of successes} = 725$$

Set the hypothesis that success and failure are in equal proportion.

Consider,

$$Z = \frac{x - \mu}{\sqrt{npq}} = \frac{725 - 670}{\sqrt{1340 \times \frac{1}{2} \times \frac{1}{2}}} = 3.005$$

where success $p = \frac{1}{2}$ and failure $q = \frac{1}{2}$ ($\because$ equal proportion).

Since calculated value of $Z = 3.005 > 2.58$, we reject the hypothesis.

Example 4: In a sample of 500 men it was found that 60% of them were overweight. What can we infer about the proportion of people being overweight in the population.

Solution: Probability of persons having overweight

$p = 60\% = 0.6$ and $q = 0.4$ $\quad (\because \quad p + q = 1)$

$$\therefore \quad \text{Probable limits} = p \pm 2.58\sqrt{\frac{pq}{n}}$$

$$= 0.6 \pm 2.58\sqrt{\frac{0.6 \times 0.4}{500}}$$

$$= 0.5435 \text{ and } 0.6565$$

$\therefore$ Probable percentage limits $= 54.35\%$ and 65.65%.

Example 5: A survey was conducted in a slum locality of 2000 families by selecting a sample of size 800. It was releaved that 180 families were illiterate. Find the probable limits of the illiterates in the population of 2000.

Solution: Probability of illiterate families

$$p = \frac{180}{800} = 0.225 \quad \text{and} \quad q = 1 - p = 0.775$$

$$\therefore \quad \text{Probable limits} = p \pm 2.58\sqrt{\frac{pq}{n}}$$

$$= 0.225 + 2.58\sqrt{\frac{0.225 \times 0.775}{800}}$$

$$= 0.225 \pm 0.038$$

$$= 0.187 \text{ and } 0.263$$

$\therefore$ The probable limits of illiterate families in a sample of 2000 is

$$= (0.187) \times 2000 \text{ and } (0.263) \times 2000$$

$$= 374 \text{ and } 526.$$

Example 6: Find how many heads in 64 tosses of a coin will ensure its fairness at 0.05 level of significance.

Solution: Let X denote the number of heads showing $N = 64$ tosses. Then distribution of X is a binomial distribution with probability of getting head in a toss is $p = \frac{1}{2}$ under the hypothesis that the coin is fair.

Here mean $= \mu = Np = 64 \times \frac{1}{2} = 32$

and $\sigma = \sqrt{npq} = \sqrt{64 \times \frac{1}{2} \times \frac{1}{2}} = 4.$

$\therefore$ The standard normal variate

$$Z = \frac{x - \mu}{\sigma} = \frac{x - 32}{4}$$

For 0.05 level of significance $Z_c = 1.96$

$$\therefore \quad -1.96 < \frac{x\ 32}{4} < 1.96$$

$$32 - 4 \times 1.96 < x < 32 + 4 \times 1.96$$

$$\therefore \quad 24.16 < x < 39.84.$$

Hence, at 0.05 level of significance, the hypothesis that the coin is fair cannot be rejected if the number of heads in 64 tosses is between 25 and 39.

Example 7: A die was thrown 1200 times and the number 6 was obtained 236 times. Can the die be considered fair at 0.01 level of significance.

Solution: Let us take the hypothesis that the die be fair at 0.01 level of significance.

The probability of getting 6 in a throw is $p = \frac{1}{6}$.

Let X be the number of 6's obtained in $N = 1200$ trials the mean of the distribution of X which is the binomial distribution.

$$\therefore \quad \mu = np = 1200 \times \frac{1}{6} = 200$$

and $$SD = \sigma = \sqrt{N \times p \times q} = \sqrt{1200 \times \frac{1}{6} \times \frac{5}{6}} = 12.91.$$

$$\therefore \quad Z = \frac{x - \mu}{\sigma} = \frac{236 - 200}{12.91} = 2.79.$$

For 0.01 level of significance we can have $Z_c = 2.58$.

Since, calculated value of $Z = 2.79$ is greater than $Z_c = 2.58$, the hypothesis has to be rejected on the basis of two tailed test.

Thus at 0.01 level of significance the die cannot be regarded as a fair one.

Example 8: The manufacturer of a certain medicine claimed that it was 90% effective in relieving an allergy for a period of 8 hours. In a sample of 200 persons who had the allergy the medicine provided relief for 160 persons.

Determine whether the manufacturer's claim is legitimate at 0.01 level of significance.

Solution: According to the manufacturer's claim $p = 0.9$

We set the hypothesis that $p = 0.9$ is correct.

Let p denote the probability that a person gets relief from allergy by using medicine and x denote the number of persons who got relief in the sample of $N = 200$ persons by using the medicine.

$$\therefore \quad p = 0.9 \quad q = 1 - p = 0.1 \quad N = 200 \quad \text{and} \quad x = 160$$

$$\therefore \quad \mu = Np = 200 \times 0.9 = 180$$

and, $$\sigma = \sqrt{Npq} = \sqrt{200 \times .9 \times .1} = 4.23$$

$$\therefore \quad Z = \frac{x-\mu}{\sigma} = \frac{160-180}{4.23} = -4.73$$

Since calculated value $|Z| = |-4.73|$ is greater than $Z_c = 2.58$, the hypothesis has to be rejected.

Example 9: The mean life-time of a sample of 100 TV tubes manufactured by a company is found to be 1580 hours with SD of 90 hours. The company claims that the mean life-time of the tubes produced by it is 1600 hours. Is the claim of the company acceptable at 0.05 level of significance? What about at 0.01 level of significance?

Solution: Consider the hypothesis $H : \mu = 1600$ hours.

The sample size is given as $N = 100$

The sample mean is $\bar{x} = 1580$ hours

and sample SD is $S = 90$ hours.

$$\therefore \quad Z = \frac{x-\mu}{\sigma} = \frac{1580-1600}{90} = \frac{-20}{90} = -0.22$$

Since $|Z| < 1.96$ and $|Z| < 2.58$.

The hypothesis should by acceptable at 5% and 1% level of significance.

Example 10: A die was thrown 9000 times and a throw of 4 and 5 was obtained 3240 times. On the assumption of random throwing, do the data indicate an unbiased die?

Solution: Let us suppose the hypothesis that the die is unbiased.

Then the probability of throwing 4 or 5 with one die $= \frac{1}{6} + \frac{1}{6} = \frac{1}{3}$.

$\therefore$ The expected number of successes $= np = 9000 \times \frac{1}{3} = 3000.$

and the observed number of successes $= 3240$

Also SD of simple sampling $= \sqrt{npq} = \sqrt{9000 \times \frac{1}{3} \times \frac{2}{3}} = 44.72$

$$\text{Hence,} \quad Z = \frac{x-np}{\sqrt{npq}} = \frac{3240-3000}{44.72} = 5.4 \text{ nearly.}$$

As $Z = 5.4 > 2.58$, the hypothesis has to be rejected at 1% level of significance and we conclude that the die is biased.

Example 11: In 324 throws of a six faced dice, odd points appeared 180 times, would you say that the die is fair at 5% level of significance.

Solution: Let us take the hypothesis that the dice is fair.

In a fair dice we except $= np = 324 \times \frac{1}{2} = 162$ odd points in 324 throws.

$$\therefore \quad SE = \sqrt{npq} = \sqrt{324 \times \frac{1}{2} \times \frac{1}{2}} = 9$$

$$\therefore \quad Z = \frac{x - np}{SE} = \frac{180 - 162}{9} = 2 > 1.96$$

Since the difference is more than 1.96 at 5% level of significance, the hypothesis is rejected. Hence, we cannot say that the dice is fair at 5% level of significance.

Example 12: In a hospital 480 female and 520 male babies were born in a week. Do these figures confirm the hypothesis that makes and females are born in equal numbers?

Solution: Let us take the hypothesis that male and female babies are born in equal numbers. i.e., $p = \frac{1}{2}$, $q = \frac{1}{2}$.

$$\therefore \quad Z = \frac{x - np}{\sqrt{npq}} = \frac{520 - 500}{\sqrt{1000 \times \frac{1}{2} \times \frac{1}{2}}} = 1.265$$

Since the difference is less than 1.95 SE (5% level).

It can be concluded that the male and female babies are born in equal numbers.

Example 13: A sample of 900 members is found to have a mean of 3.4 cm can it be reasonably regarded as a truly random sample from a large population with mean 3.25 cm and SD 1.61 cm.

Solution: Given $\bar{x} = 3.4$ cm $\quad n = 900$

$$\mu = 3.25 \text{ cm and } \sigma p = 1.61 \text{ cm.}$$

Consider, $Z = \dfrac{\bar{x} - \mu}{\sigma/\sqrt{n}} = \dfrac{3.4 - 3.25}{1.61/\sqrt{900}} = 2.8$

As $Z = 2.8 > 1.96$, the deviation of the sample mean from the mean of the population is significant at 5% level of significance. Hence, it cannot be regarded as a random sample.

8.2.3 Test for Proportion of Successes

Instead of recording the number of successes in each sample, we may record the proportion success. i.e., $\frac{1}{n}$ th of the number of successes in each sample. As this would amount to dividing all figures of the record by n, the mean proportion of successes must be p and the S.D. of the proportion of successes $\sqrt{\frac{pq}{n}}$. Thus we may have the following formula.

$$SE_p = \sqrt{\frac{pq}{n}}$$

Example 1: 500 apples are taken at random from a large basket and 50 are found to be bad. Estimate the proportion of bad apples in the basket and assign limits within which the % must probably lies?

Solution: The proportion of bad apples in the given

$$\text{apples} = \frac{50}{500} = 0.1 \quad \therefore \quad p = 0.1 \quad \text{and} \quad q = 0.9.$$

$$\therefore \quad SE_p = \sqrt{\frac{pq}{n}} = \sqrt{\frac{.1 \times .9}{500}} = 0.013.$$

The limits within which percentage of bad apples lies $\left| p \pm 3\sqrt{\frac{pq}{n}} \right| \times 100.$

$$= |0.1 \pm 3(0.013)| \times 100$$
$$= 6.1 \text{ and } 13.9.$$

Thus, the percentage of bad apples in the consignment almost certainly lies between 6.1 and 13.9.

Example 2: A wholesaler in apples claims that only 4% of the apples supplied by him are defective. A random sample of 600 apples contained 36 defective apples. Test the claim of the wholesaler.

Solution: The wholesaler claims that 4% of the apples are defective.

Hence, the 95% of the confidence limits are given by

$$p \pm 1.96\, SE$$

Now, $p = 0.96$ i.e., the probability of good apples and

$q = 0.04$ probability of defective apples.

$$\therefore \quad SE_P = \sqrt{\frac{pq}{n}} = \sqrt{\frac{.96 \times 0.04}{600}} = 0.008$$

$\therefore$ 95% confidence limits are

$$p \pm 1.96\, SE$$

$$= 0.96 \pm 1.96 \times 0.008$$

$$= 0.96 \pm 0.0157$$

$$= 0.9443 \text{ to } 0.9757$$

Out of 600 apples the good apples may vary between $0.9443 \times 600 = 566.50$ and $0.9757 \times 600 = 585.42$

i.e., 567 and 585.

The number of defectives is thus expected to lie between 15 and 33. Since the actual number is 36, the wholesaler's claim that only 4% of the apples are defective cannot be accepted.

8.2.4 Comparison of Large Samples

Two large samples of size n_1, n_2 are taken from two populations giving proportions of attributes A's as P_1 and P_2 respectively.

(a) On the hypothesis that the populations are similar as regards the attributes A.

We combine the two samples to find an estimate of the common value of proportions of A's in the populations which is given by

$$p = \frac{n_1 p_1 + n_2 p_2}{n_1 + n_2} \quad \text{or} \quad p = \frac{x_1 + x_2}{n_1 + n_2}$$

If e_1, e_2 be the standard errors in the two samples then $e_1^2 = \frac{pq}{n_1}$ and $e_2^2 = \frac{pq}{n_2}$

If e be the standard error of the difference between p_1 and p_2 then

$$e^2 = e_1^2 + e_2^2 = pq\left(\frac{1}{n_1} + \frac{1}{n_2}\right)$$

$$\therefore \quad Z = \frac{p_1 - p_2}{e}$$

If $Z > 3$, the difference between p_1 and p_2 is real one.
If $Z < 2$, the difference may be due to fluctuations of simple sampling.
But if Z lies between 2 and 3 then the difference is significant at 5% level of significance.

(b) If the proportions of A's are not the same in the two populations from which the samples are drawn but p_1 and p_2 are the true values of proportions then *SE*'e' of the difference $p_1 - p_2$ is given by

$$e^2 = \frac{p_1 q_1}{n_1} + \frac{p_2 q_2}{n_2}$$

If $Z = \frac{p_1 - p_2}{e} < 3$, the difference could be due to fluctuations of simple sampling.

Example 1: In a city A, 20% of a random sample of 900 school boys had a certain slight physical defect. In another city B, 18.5% of a random sample of 1600 school boys had the same defect. Is the difference between the proportions significant?

Solution: Given $n_1 = 900$ and $n_2 = 1600$

$$p_1 = \frac{20}{100} = \frac{1}{5} \quad \text{and} \quad p_2 = \frac{18.5}{100}$$

$$\therefore \quad p = \frac{n_1 p_1 + n_2 p_2}{n_1 + n_2} = \frac{180 + 296}{900 + 1600} = 0.19$$

and $q = 1 - p = 1 - 0.19 = 0.81$

$$\therefore \quad e^2 = pq\left(\frac{1}{n_1} + \frac{1}{n_2}\right) = 0.19 \times 0.81\left(\frac{1}{900} + \frac{1}{1600}\right)$$

$$= 0.0017$$

$$\therefore \quad e = \sqrt{0.0017} = 0.04 \quad \text{nearly.}$$

$$\therefore \quad Z = \frac{p_1 - p_2}{e} = \frac{\frac{20}{100} - \frac{18.5}{100}}{0.04} = \frac{0.015}{0.04} = 0.37.$$

As $Z < 1$, the difference between the proportions is significant.

Example 2: In two large populations, there are 30% and 25% respectively of fair haired people. Is this difference likely to be hidden in samples of 1200 and 900 respectively from the two populations.

Solution: Given $n_1 = 1200$, $n_2 = 900$

$$p_1 = 0.3 \quad \text{and} \quad p_2 = 0.25$$

$$\text{Now,} \quad e^2 = \frac{p_1 q_1}{n_1} + \frac{p_2 q_2}{n_2} = \frac{0.3 \times 0.7}{1200} + \frac{0.25 \times 0.75}{900}$$

$$\therefore \quad e = 0.0195$$

$$\therefore \quad Z = \frac{p_1 - p_2}{e} = \frac{0.3 - 0.25}{0.0195} = 2.5 \quad \text{nearly.}$$

Hence, it is unlikely that the real difference will be hidden.

8.2.5 Sampling Distribution of the Mean

If a population is distributed normally with mean μ and SD σ, then the sample means μ_1, $\mu_2 \ldots$ are also distributed normally with mean μ and standard error $\frac{\sigma}{\sqrt{n}}$. This result shows how the precision of a sample mean increases as the sample size increases.

8.2.6 Central Limit Theorem

This is an important theorem regarding the distribution of the mean of a sample if the parent population is non-normal and sample size is large.

If the variable X has a non-normal distribution with mean μ and SD = σ then the limiting distribution of $Z = \dfrac{\bar{x} - \mu}{\sigma/\sqrt{n}}$ as $n \to \infty$ is the standard normal distribution (i.e., mean is 0 and SD is 1).

This theorem holds good for a sample of 25 or more which is regarded as large.

Confidence limits for the unknown mean:

If $\bar{X}$ is not significant at 5% level of probability then

$$\left|(\bar{X}-\mu)\frac{\sqrt{n}}{\sigma}\right| < 1.96.$$

$$\text{i.e.,}\quad \bar{X}-1.96\frac{\sigma}{\sqrt{n}} < \mu < \bar{X}+1.96\frac{\sigma}{\sqrt{n}}$$

Thus 95% confidence limits for population mean μ are $\bar{X}\pm 1.96\frac{\sigma}{\sqrt{n}}$.

Similarly,

$$99\% \text{ confidence limits for } \mu \text{ are } \bar{X}\pm 2.58\frac{\sigma}{\sqrt{n}}.$$

Example 1: Before an increase in excise duty on tea 400 people out of a sample of 500 person's were found to be tea drinkers. After an increase in duty, 400 persons were known to be tea drinkers in a sample of 600 people. Do you think that there has been a significant decrease in the consumption of tea after the increase in the excise duty.

Solution: Let us take the hypothesis that there is no significant decrease in the consumption of tea after increase in the excise duty.

The *SE* of the difference of proportions

$$e = SE(p_1 - p_2) = \sqrt{pq\left(\frac{1}{n_1}+\frac{1}{n_2}\right)}$$

$$p_1 = \frac{400}{500} = 0.8, \quad p_2 = \frac{400}{600} = 0.667$$

$$p = \frac{400+400}{500+600} = \frac{8}{11} \quad \therefore \quad q = 1 - p = \frac{3}{11}.$$

$$\therefore \quad SE(p_1 - p_2) = \sqrt{\frac{8}{11}\times\frac{3}{11}\left(\frac{1}{500}+\frac{1}{600}\right)} = 0.027.$$

$$\therefore \quad Z = \frac{p_1 - p_2}{SE} = \frac{0.8 - 0.67}{0.027} = 4.93$$

Since the difference is more than 2.58 (SE) (1% level of significance) it could not have arisen due to fluctuations of sampling. Hence, there is a significant decrease in the consumption of tea after the increase in excise duty.

Example 2: A machine produced 20 defective articles in a batch of 400. After overhauling it produced 10 defectives in a batch of 300. Has the machine improved.

Solution: Let us take the hypothesis that the machine has not improved after overhauling.

$$\text{Here,} \quad p_1 = \frac{20}{400} = 0.050, \quad p_2 = \frac{10}{300} = 0.033.$$

$$p = \frac{x_1 + x_2}{n_1 + n_2} = \frac{20 + 10}{400 + 300} = 0.043. \quad \text{and} \quad q = 1 - p = 0.957.$$

$$SE(p_1 - p_2) = \sqrt{pq\left(\frac{1}{n_1} + \frac{1}{n_2}\right)} = \sqrt{0.043 \times 0.957\left(\frac{1}{400} + \frac{1}{300}\right)} = 0.0155.$$

$$\therefore \quad Z = \frac{p_1 - p_2}{SE} = \frac{0.05 - 0.033}{0.0155} = 1.1$$

Since the difference is less than 1.96 *SE* (at 5% level) our hypothesis is true, the machine has not improved after overhauling.

Example 3: In a simple random sample of 600 men taken from a big city 400 are found to be smokers. In another simple random sample of 900 men taken from another city 450 are smokers. Do the data indicate there is a significant difference in the habit of smoking in the two cities?

Solution: Let us take the hypothesis that there is no significant difference in the habit of smoking in the two cities.

$$\text{Here} \quad p_1 = \frac{400}{600} = 0.667 \quad \text{and} \quad p_2 = \frac{450}{900} = 0.5$$

$$\text{where} \quad n_1 = 600, \ n_2 = 900 \ X_1 = 400 \quad \text{and} \quad X_2 = 450$$

$$\text{Now,} \quad p = \frac{X_1 + X_2}{n_1 + n_2} = \frac{400 + 450}{600 + 900} = \frac{17}{30} \quad \text{and} \quad q = 1 - p = \frac{13}{30}$$

$$\text{and} \quad SE(p_1 - p_2) = \sqrt{pq\left(\frac{1}{n_1} + \frac{1}{n_2}\right)}$$

$$= \sqrt{\frac{17}{30} \times \frac{13}{30}\left(\frac{1}{600} + \frac{1}{900}\right)} = 0.026$$

$$\therefore \quad Z = \frac{p_1 - p_2}{SE} = \frac{0.667 - 0.5}{0.026} = 6.42.$$

Since the difference is more than 2.58 *SE* at 1% level of significance the hypothesis is rejected. Hence, there is a significant difference in the habit of smoking of the two cities.

8.2.7 Test of Significance for Means of Two Large Samples

Suppose two random samples of size n_1 and n_2 have been drawn from the same population with *SD* σ, we wish to test whether the difference between the sample means $\bar{x}_1$ and $\bar{x}_2$ is significant or is merely due to fluctuations of sampling.

If the samples are independent, then the $SE(e)$ of the difference of their means is given

$$e^2 = e_1^2 + e_2^2$$

where $e_1 = \frac{\sigma}{\sqrt{n_1}}$ and $e_2 = \frac{\sigma}{\sqrt{n_2}}$ are the SES of the means of the two samples.

and $e = \sigma\sqrt{\frac{1}{n_1} + \frac{1}{n}}$

$$\therefore \quad Z = \frac{\bar{x}_1 - \bar{x}_2}{\sigma\sqrt{\frac{1}{n_1} + \frac{1}{n_2}}}$$ is normally distributed with mean 0 and SD is 1.

If $Z > 1.96$ then the difference is significant at 5% level of significance.

If $Z > 3$, it is highly probable that either the samples have not been drawn from the same population.

If the samples are known to be drawn from different populations with means μ_1, μ_2 and SDS σ_1 and σ_2. Then $SE(e)$ of their means is given by

$$e = \sqrt{\frac{\sigma_1^2}{n_1} + \frac{\sigma_2^2}{n_2}}.$$

Example 1: The means of simple samples of size 1000 and 2000 are 67.5 cm and 68 cm respectively. Can the samples be regarded as drawn from the same population of SD 2.5 cm.

Solution: Given $n_1 = 1000$, $n_2 = 2000$

$$\bar{x}_1 = 67.5 \quad \bar{x}_2 = 68.0 \quad \text{and} \quad \sigma = 2.5 \text{ cm.}$$

on the hypothesis that the samples are drawn from the population of $SD = \sigma = 2.5$

$$Z = \frac{\bar{x}_1 - \bar{x}_2}{\sigma\sqrt{\frac{1}{n_1} + \frac{1}{n_2}}}$$

$$= \frac{67.5 - 68.0}{2.5\sqrt{\frac{1}{1000} + \frac{1}{2000}}} = |-5.1| = 5.1.$$

Hence, the difference between the sample means i.e., 5.1 is very much greater than 1.96 and is therefore significant. Thus the samples cannot be regarded as drawn from the same population.

8.2.8 Standard Error of Mean

(i) When standard deviation of the population is known

$$SE\ \bar{X} = \frac{\sigma p}{\sqrt{n}}.$$

where $SE\ \bar{X}$ refers to the SE of the mean

σp – standard deviation of the population

n – number of observations in the sample.

(ii) When standard deviation of population is not known we have to use standard deviation of the sample in calculating error of mean.

$$SE\ \bar{X} = \frac{\sigma \text{ (sample)}}{\sqrt{n}}$$

where σ denotes the SD of the sample.

Example 1: Calculate standard error of the mean from the following data showing the amount paid by 100 firms in Kolkata on the occation of Durga Pooja.

Mid value (Rs):	39	49	59	69	79	89	99
No of firms:	2	3	11	20	32	25	7

Solution: First we have to find $SE\ \bar{X} = \dfrac{\sigma}{\sqrt{n}}$.

$$\text{where } \sigma = \sqrt{\frac{\sum d^2 f}{\mu} - \left(\frac{\sum fd}{\mu}\right)^2} \times h$$

$$N = 100$$

μvm	f	$d = \dfrac{m-69}{10}$	fd	fd^2
39	2	– 3	– 6	18
49	3	– 2	– 6	12
59	11	– 1	– 11	11
69	20	0	0	0
79	32	1	32	32
89	25	2	50	100
99	7	3	21	63
	$N = 100$		$\sum fd = 80$	$\sum fd^2 = 236$

$$\therefore \quad \sigma = \sqrt{\frac{236}{100} - \left(\frac{80}{100}\right)^2} \times 10 = 13.11$$

$$SE\ \bar{X} = \frac{13.11}{\sqrt{100}} = 1.311.$$

Example 2: The mean height obtained from a random sample of size 100 is 64 inches. The SD of the distribution of height of the population is known to be 3 inches. Test the statement that the mean height of the population is 67 inches at 5% level of significance. Also set of 99% of limits of the mean height of the population.

Solution: Let us take the hypothesis that there is no significant difference between the sample mean and the population mean.

$$SE\ \bar{X} = \frac{\sigma}{\sqrt{n}} = \frac{3}{\sqrt{100}} = 0.3$$

$$\frac{\bar{X} - \mu}{SE} = \frac{67 - 64}{0.3} = 10.$$

Since the difference is more than 1.96 *SE* (5% level) the hypothesis is rejected. Hence, the mean height of the population could not be 67 inches.

Now, 99% probable limits of the mean height of the population

$$= \bar{X} \pm 2.58\, SE$$

$$= 64 \pm 2.58(.3)$$

$$= 64 \pm 0.774$$

$$= 63.2 \quad \text{to} \quad 64.8.$$

Example 3: A sample of 100 tyres is taken from a lot. The mean life of tyres is found to be 39350 km with SD of 3260. Could the sample come from a population with mean life of 40,000 km? Establish 99% confidence limits within which the mean life of tyres is expected to lie.

Solution: Let us take the hypothesis that there is no significant difference between the sample mean and the hypothetical population mean

$$SE\,\bar{X} = \frac{\sigma}{\sqrt{n}} = \frac{3260}{\sqrt{100}} = 326$$

$$\therefore \quad \frac{\text{Diff}}{SE\,\bar{X}} = \frac{40,000 - 39350}{326} = 1.994.$$

Since the difference is less than 2.58 *SE* (at 1% level) the hypothesis is accepted and the difference could have arisen due to fluctuations of sampling.

$$\text{Now,} \quad 99\% \text{ confidence limits} = \bar{X} \pm 2.58\, SE$$

$$= 39350 + 2.58 \times 326$$

$$= 39350 \pm 841.08$$

Hence, the mean life of tyres is expected to lie between 38509 and 40191.

8.2.9 Chi-Square Distribution (χ^2 – test)

Chi-square distribution provides a measure of correspondence between theoretical and observed frequencies. If O_i $(i = 1,2,3\ldots n)$ and E_i $(i = 1,2,3\ldots n)$ respectively denotes a set of observed and estimated frequencies, the quantity Chi-square denoted by χ^2 is defined as

$$\chi^2 = \sum \frac{(O_i - E_i)^2}{E_i}$$

with degrees of freedom $(n-1)$.

8.2.10 Chi-square Test as a Test of Goodness of Fit

Chi-square test is a non-parametric test and is used as a test of goodness of fit. χ^2-test enables us to see how well does the assumed theoretical distribution (such as binomial, Poisson or normal distributions) fit to the observed data. When some theoretical distribution is fitted to the given data, we are always interested in knowing as to how well this distribution fits with the observed data. The Chi-square test can give answer to this. If the calculated value of χ^2 is less than the table value at a certain level of significance, the fit is considered to be a good one which means that the divergence between the observed and expected frequencies is attributable to fluctuations of sampling. But if the calculated value of χ^2 is greater than its table value, the fit is not considered to be good one.

Also, as a test of independence, χ^2 test enables us to explain whether or not two attributes are associated.

Example 1: For the following data test the hypothesis that the accidents are uniformly distributed over all the days of the week for 99% confidence.

Day:	Sun	Mon	Tue	Wed	Thur	Fri	Sat	Total
No of accidents:	14	16	8	12	11	9	14	84

Solution: First set the hypothesis (H_o) that the accidents are uniformly distributed over all the days of a week.

The frequencies in the given data are observed frequencies.

The expected frequencies for the days are $\frac{84}{7} = 12$ each day.

Now, the data is as follows:

Days:	Sun	Mon	Tue	Wed	Thur	Fri	Sat
O_i	14	16	8	12	11	9	14
E_i	12	12	12	12	12	12	12

$$\therefore \quad \chi^2 = \sum \frac{(O_i - E_i)^2}{E_i}$$

$$= \frac{(14-12)^2}{12} + \frac{(16-12)^2}{12} + \frac{(8-12)^2}{12} + \frac{(12-12)^2}{12} + \frac{(11-12)^2}{12}$$

$$+ \frac{(9-12)^2}{12} + \frac{(14-12)^2}{12}$$

$$= \frac{4+16+16+0+1+9+4}{12} = \frac{50}{12} = 4.25.$$

Table value of $\chi^2 = 16.81$ for $v = 6$ d.f. 1% level of significance. Calculated value of χ^2 is less than the table value of χ^2 we accept the hypothesis that the accidents are uniformly distributed over all the days of the week with 99% confidence.

Example 2: A die is thrown 264 times and the number appearing on the face (x) follows the following distribution

x:	1	2	3	4	5	6	Total
$f(x)$:	40	32	28	50	54	60	264

Calculate the value of χ^2.

Solution: The frequencies in the given data are the observed frequencies for the numbers 1, 2, 3, 4, 5, 6

The expected number of frequencies for the numbers 1, 2, 3, 4, 5, 6 to appear on the face is $= \frac{264}{6} = 44$ each.

Now, the data is as follows:

No of the die:	1	2	3	4	5	6
O_i:	40	32	28	50	54	60
E_i:	44	44	44	44	44	44

$$\therefore \quad \chi^2 = \sum \frac{(O_i - E_i)^2}{E_i}$$

$$= \frac{(40-44)^2}{44} + \frac{(32-44)^2}{44} + \frac{(28-44)^2}{44} + \frac{(50-44)^2}{44} + \frac{(54-44)^2}{44} + \frac{(60-44)^2}{44}$$

$$= \frac{968}{44} = 22.$$

Example 3: Five dice were thrown 96 times and the numbers 1, 2 or 3 appearing on the face of the die follows the frequency distribution as follows:

No of dice showing: 1, 2, 3	5	4	3	2	1	0
Frequency:	7	19	35	24	8	3

Test the hypothesis that the data follows a binomial distribution (given $\chi^2_{0.05} = 11.07$ for 5 d.f).

Solution: The data gives the observed frequencies and we need to calculate the expected frequencies.

Probability of a single die throwing 1, 2 or 3 is $p = \frac{3}{6} = \frac{1}{2}$ $p + q = 1$

$\therefore \quad q = \frac{1}{2}$

The binomial distribution of fit is

$$N(q+p)^n = 96\left(\frac{1}{2}+\frac{1}{2}\right)^5$$

The theoretical frequencies of getting 5, 4, 3, 2, 1, 0 successes with 5 dice are respectively the successive terms of the binomial expansion.

They are:

$$96 \times \frac{1}{2^5},\ 96\ {}^5C_1\frac{1}{2^5},\ 96\ {}^5C_2\frac{1}{2^5},\ 96\ {}^5C_3\frac{1}{2^5},\ 96\ {}^5C_4\frac{1}{2^5} \quad \text{and} \quad 96\ {}^5C_5\frac{1}{2^5}.$$

$= 3,\ 15,\ 30,\ 30,\ 15,\ 3$

Thus, we have the table of observed and expected frequencies.

O_i:	7	19	35	24	8	3
E_i:	3	15	30	30	15	3

$$\therefore \quad \chi^2 = \sum \frac{(O_i - E_i)^2}{E_i}$$

$$= \frac{(7-3)^2}{3} + \frac{(19-15)^2}{15} + \frac{(35-30)^2}{30} + \frac{(24-30)^2}{30} + \frac{(8-15)^2}{15} + \frac{(3-3)^2}{3}$$

$$= 11.7$$

Table value of $\chi^2_{0.05} = 11.07$ for 5 d.f.

The calculated value of χ^2 is greater than the table value of χ^2, we reject the hypothesis that the data follows a binomial distribution.

Example 4: A sample analysis of examination results of 500 students was made. It was found that 220 students had failed, 170 had secured third class, 90 had secured second class and 20 had secured firs class. Do these figures support the general examination result which is in the ratio 4:3:2:1 for the respective categories.

(Given $\chi^2_{0.05} = 7.81$ for 3 d.f)

Solution: Let us take the hypothesis that these figures support to the general result in the ratio 4 : 3 : 2 : 1. The expected frequencies in the respective categories are:

$$\frac{4}{10} \times 500 = 200, \quad \frac{3}{10} \times 500 = 150, \quad \frac{2}{10} \times 500 = 100, \quad \frac{1}{10} \times 500 = 50$$

Thus, we have

O_i :	220	170	90	20
E_i :	200	150	100	50

$$\therefore \quad \chi^2 = \frac{\Sigma(O_i - E_i)^2}{E_i}$$

$$= \frac{(220-200)^2}{200} + \frac{(170-150)^2}{150} + \frac{(90-100)^2}{100} + \frac{(20-50)^2}{50}$$

$$= 23.67$$

Since the calculated value of $\chi^2 = 23.67$ is greater than the table value of $\chi^2_{0.05} = 7.81$, the hypothesis is rejected.

Example 5: Four coins are tossed 100 times and the following results were obtained. Fit a binomial distribution for the data and test the goodness of fit.

(Given $\chi^2_{0.05} = 9.49$ for 4 d.f)

No of heads:	0	1	2	3	4
Frequency:	5	29	36	25	5

Solution: From the given data we calculate the mean as

$$\mu = \frac{\Sigma x \times f}{\Sigma f} = \frac{196}{100} = 1.96$$

For binomial distribution $\mu = np$

$$\therefore \quad 1.96 = 4p \quad \therefore \quad p = 0.49 \quad \text{and} \quad q = 1 - p = 0.51.$$

Now, the binomial distribution of fit is

$$N(q+p)^n = 100\left(q^4 + {}^4C_1 q^3 p + {}^4C_2 q^2 p^2 + {}^4C_3 q p^3 + p^4\right)$$

$$= 100\left[(.51)^4 + 4(.51)^3(.49) + 6(.51)^2(.49)^2 + 4(.51)(.49)^3 + (.49)^4\right]$$

$$= 7,\ 26,\ 37,\ 24,\ 6$$

Thus					
O_i:	5	29	36	25	5
E_i	7	26	37	24	6

$$\therefore \quad \chi^2 = \frac{\Sigma(O_i - E_i)^2}{E_i} = 1.15$$

Table value of $\chi^2_{0.05} = 9.49$ for 4 d.f.

Since the calculated value of χ^2 is less than the table value of χ^2, the hypothesis that the fitness is good can be accepted.

Example 6: Fit a Poisson distribution for the following data and test the goodness of fit, given that $\chi^2 = 0.05 = 7.815$ for 3 d.f.

x:	0	1	2	3	4
f:	122	60	15	2	1

Solution: First we calculate the mean from the given data

$$\text{mean, } \mu = \frac{\Sigma x \times f}{\Sigma f} = \frac{100}{200} = 0.5.$$

For Poisson distribution $\bar{x} = \mu$ and the Poisson distribution of fit is

$$f(x) = N\frac{(e^{-\mu}\mu^x)}{x!} : \quad x = 0,\ 1,\ 2,\ 3,\ 4.$$

Putting $x = 0, 1, 2, 3, 4$ respectively we get the theoretical frequencies.

Thus	121,	61,	15,	3,	0
$\therefore$ O_i :	122	60	15	$2 + 1 = 3$	
E_i :	121	61	15	$3 + 0 = 3$	

$$\therefore \quad \chi^2 = \sum \frac{(O_i - E_i)^2}{E_i} = \frac{(122 - 121)^2}{121} + \frac{(60 - 61)^2}{61} + 0 + 0$$

$$= 0.025$$

Thus, the calculated value of $\chi^2 = 0.025$ is less than the table value of $\chi^2_{0.05} = 7.815$ for 3 d.f.

$\therefore$ The fitness is considered to be good. i.e., the hypothesis that the fitness is good can be accepted.

Example 7: The number of accidents per day (x) as recorded in a textile industry over a period of 400 days is given below. Test the goodness of fit in respect of Poisson distribution of fit to the given data.

x :	0	1	2	3	4	5
f :	173	168	37	18	3	1

Given $\chi^2_{0.05} = 9.49$ for 4 d.f.

Solution: To fit a Poisson distribution we calculate the mean as

$$\text{Mean} = \mu = \frac{\Sigma x \times f}{\Sigma f} = \frac{313}{400} = 0.7825$$

The Poisson distribution for fit is

$$f(x) = N\left(\frac{e^{-\mu}\mu^x}{x!}\right) = 400\left(\frac{e^{-0.7825}(0.7825)^x}{x!}\right)$$

Putting $x = 0, 1, 2, 3, 4, 5$ respectively we get 183, 143, 56, 15, 3, 0

Thus O_i :	173	168	37	18	$3 + 1 = 4$
E_i :	183	143	56	15	$3 + 0 = 3$

$$\therefore \quad \chi^2 = \frac{\Sigma(O_i - E_i)^2}{E_i} = \frac{(173-183)^2}{183} + \frac{(168-143)^2}{143} + \frac{(37-56)^2}{56} + \frac{(18-15)^2}{15} + \frac{(4-3)^2}{3}$$

$$= 12.3.$$

Thus, calculate $\chi^2 = 12.3 >$ table value of $\chi^2_{0.05} = 9.49$.

We conclude that the fitness is not good and we reject the hypothesis.

8.2.11 Paired *t*-test

(Hypothesis testing for comparing two related samples)

Paired '*t*' test is a way to test for comparing two related samples, involving small values of *n* that does not require the variances of the two populations to be equal but the assumption that the two populations are normal must continue to apply.

To apply this test:

(i) We first work out the difference score for each matched pair.

(ii) Find out the average of such differences $\bar{D}$

$$\text{i.e.,} \quad \bar{D} = \frac{\Sigma D_i}{n}$$

(iii) Calculate the variance of the differences

$$\text{i.e.,} \quad (\sigma_{\text{diff}})^2 = \frac{\Sigma D_i^2 - (\bar{D})^2 \cdot n}{n-1}$$

(iv) Assuming the said differences to be normally distributed and independent we work out *t* as

$$t = \frac{\bar{D} - 0}{\sigma_{\text{diff}} / \sqrt{n}} \quad \text{with } (n-1) \text{ degrees of freedom.}$$

Where *n* is the number of matched pairs.

Example 1: Eleven students were given a test in mathematics. They were given a month's further tuition and a second test of equal difficulty was held at the end of it. Do the marks give evidence that the students have benefited by extra coaching?

Students:	1	2	3	4	5	6	7	8	9	10	11
M.I test:	23	20	19	21	18	20	18	17	23	16	19
M.II test:	24	19	22	18	20	22	20	20	23	20	17

Solution: Let us assume the hypothesis that the students have not been benefited by extra coaching. It implies that the mean of the difference between marks of the two tests is zero i.e., $\mu = \bar{D} = 0$.

To compute the mean and SD of the difference between the marks of the two tests we form the table as below:

Students	x	y	$D = y - x$	$D - \bar{D}$	$(D - \bar{D})^2$
1	23	24	1	0	0
2	20	19	– 1	– 2	4
3	19	22	3	2	4
4	21	18	– 3	– 4	16
5	18	20	2	1	1
6	20	22	2	1	1
7	18	20	2	1	1
8	17	20	3	2	4
9	23	23	0	– 1	1
10	16	20	4	3	9
11	19	17	– 2	– 3	9
			$\Sigma D = 11$		$\Sigma(D - \bar{D})^2 = 50$

Now,

$$\bar{D} = \text{Mean of differences} = \frac{\Sigma D}{n} = \frac{11}{11} = 1$$

$$\sigma_{\text{diff}}^2 = \frac{\Sigma(D - \bar{D})^2}{n - 1} = \frac{50}{10} = 5$$

$$\therefore \quad \sigma_{\text{diff}} = 2.24$$

$$\therefore \quad t = \frac{\bar{D} - \mu}{\sigma_{\text{diff}}}\sqrt{n}$$

$$= \frac{(1 - 0)}{2.24}\sqrt{11} = 1.48 \text{ nearly.}$$

From the table $t_{0.05} = 2.228$ for 10 d.f.

As the calculated value of $t <$ the table value of $t_{0.05}$, the value of t is not significant at 5% level of significance, i.e., the test provides no evidence that the students have benefitted by extra coaching.

Example 2: Memory capacity of 9 students was tested before and after training. State at 5% level of significance whether the training was effective from the following scores:

Students:	1	2	3	4	5	6	7	8	9
Before:	10	15	9	3	7	12	16	17	4
After:	12	17	8	5	6	11	18	20	3

Use paired t-test.

Solution: Take the score before training as X and the score after training as Y and then taking the null hypothesis that the mean of difference is zero, we can write

$$H_0 : \mu_1 = \mu_2 \text{ which is equivalent to test } H_0 : \bar{D} = 0$$
$$H_\alpha : \mu_1 < \mu_2 \text{ (training has been effective).}$$

We use the paired t-test as

$$t = \frac{\bar{D} - 0}{\sigma_{\text{diff}} / \sqrt{n}}$$

Students	***x***	***y***	$D_i = X - Y$	D_i^2
1	10	12	−2	4
2	15	17	−2	4
3	9	8	1	1
4	3	5	−2	4
5	7	6	1	1
6	12	11	1	1
7	16	18	−2	4
8	17	20	−3	9
9	4	3	1	1
$n = 9$			$\sum D_i = -7$	$\sum D_i^2 = 29$

$$\text{Mean of differences } \bar{D} = \frac{\sum D_i}{n} = \frac{-7}{9} = -0.778$$

$$\text{SD of differences } \sigma_{\text{diff}} = \sqrt{\frac{\sum D_i^2 - (\bar{D})^2 \cdot n}{n-1}}$$

$$= \sqrt{\frac{29-(-0.778)9}{n-1}} = 1.715$$

$$\therefore \quad t = \frac{-0.778-0}{1.715/\sqrt{9}} = -1.361$$

Table value $t_{0.05} = -1.86$ for 8 d.f.

Since the calculated value of $t = -1.361$ is less than the table value of $t_{0.05} = -1.86$ we accept H_0 and we conclude that the difference in score before and after training is insignificant. Hence, we say that the training was not effective.

Example 3: From a random sample of 10 pigs fed on diet A, the increases in weight in a certain period were 10, 6, 16, 17, 13, 12, 8, 14, 15, 9 lbs. For another random variable of 12 pigs fed on diet B, the increases in the same period were 7, 13, 22, 15, 12, 14, 18, 8, 21, 23, 10, 17 lbs. Test whether diets A and B differ significantly as regards their effect on increases in weight?

Solution: We calculate the means and standard deviation of the samples as follows:

Diet A			**Diet B**		
x	$x-\bar{x}$	$(x-\bar{x})^2$	y	$y-\bar{y}$	$(y-\bar{y})^2$
10	−2	4	7	−8	16
6	−6	36	13	−2	4
16	4	16	22	7	49
17	5	25	15	0	0
13	1	1	12	−3	9
12	0	0	14	−1	1
8	−4	16	18	3	9
14	2	4	8	−7	49
15	3	9	21	6	36
9	−3	9	23	8	64
120	0	120	10	−5	25
			17	2	4
			180	0	314

$$\bar{x} = \frac{120}{10} = 12\ lbs, \quad \bar{y} = \frac{180}{12} = 15\ lbs$$

$$\sigma_s^2 = \frac{\Sigma(x_i - \bar{x})^2 + \Sigma(y_i - \bar{y})^2}{n_1 + n_2 - 2}$$

$$= \frac{120 + 314}{(10 + 12 - 2)} = \frac{434}{20} = 21.1$$

$$\therefore \quad \sigma_s = 4.65.$$

Assuming that the samples do not differ in weight so far as the two diets are concerned, i.e.,

$$\mu_1 - \mu_2 = 0$$

$$\therefore \quad t = \frac{(\bar{y} - \bar{x}) - 0}{\sigma_s \sqrt{\left(\frac{1}{n_1} + \frac{1}{n_2}\right)}} = \frac{15 - 12}{4.65\sqrt{\frac{1}{10} + \frac{1}{12}}} = 1.6 \text{ nearly.}$$

Here d.f. $v = n_1 + n_2 - 2 = 20$ $t_{0.05} = 2.09$.

$\therefore$ Calculated value of $t < t_{0.05}$

Hence, the difference between the sample means is not significant, i.e., the two diets do not differ significantly as regards their effect on increase in weight.

QUESTION BANK

1. Define the following:
   (a) Type-I error and type-II error.
   (b) One-tailed test and two-tailed test.
   (c) Null hypothesis and alternative hypothesis
   (d) Level of significance and confidence limits.
   (e) Parameter and statistic.
2. In a city *A* 20% of a random sample of 900 school boys had a certain slight physical defect. In another city *B* 18.5% of a random sample of 1600 school boys had the same defect. Is the difference between the proportions significant?

   Ans: $Z = 0.37$

3. A sample of 1000 days is taken from meteorological records of a certain district and 120 of them are found to be foggy. What are the probable limits to the percentage of foggy days in the district?

   Ans: 8.91% and 15.07%

4. A machine produces 16 imperfect articles in a sample of 500. After the machine is overhauled, it produces 3 imperfect articles in a batch of 100. Has the machine been improved.

   Ans: No.

5. In a sample of 600 men from a certain city, 450 are found smokers. In another sample of 900 men from another city, 450 are smokers. Do the data indicate that the cities are significantly different with respect to the habit of smoking among men?

   Ans: $Z = 6.56$ so that the difference is significant.

6. The mean life-time of a sample of 100 fluorescent light bulbs produced by a company is computed to be 1570 hours with a standard deviation of 120 hours. If n is the life-time of all the bulbs produced by the company, test the hypothesis, $n = 1600$ hours against the alternative hypothesis $n \neq 1600$ hours using a level of significance of 0.01.

7. Fit a Poisson distribution to the following data and test for its goodness of fit at 5% level of significance.

x:	0	1	2	3	4
f:	419	352	154	56	19

   Given $\chi^2_{0.05}$ for d.f. 3 is 7.82.

8. A coin was tossed 400 times and the head turned up 216 times. Test the hypothesis that the coin is unbiased at 5% level of significance.

9. A random sample of 10 boys had the following I.Q. 70, 120, 110, 101, 88, 83, 95, 98, 107, 100. Do the data support the assumption of a population mean I.Q. of 100 at 5% level of significance?

10. A die is tossed 960 times and it falls with 5 upwards 184 times. Is the die biased at 5% level of significance?

11. In 324 throws of a six faced die, an odd number turned up 181 times. Is it reasonable to think that the die is an unbiased one?

12. A survey conducted on 64 families with 3 children each and recorded as follows:

No of male children:	0	1	2	3
No of families:	6	19	29	10

Apply χ^2-test to test whether male and female children are equiprobable at 5% level of significance.

MODEL QUESTION PAPER

Fourth Semester B.E. Degree Examination

Time: 3hrs **Max. Marks** - 100

Note: Answer any Five full questions choosing at least two from each section.

PART – A

1. (a) Use Picard's method to approximate the value of y when $x = 0.1$ given that $y = 1$ when $x = 0$ and $\frac{dy}{dx} = 3x + y^2$.

   (b) Using modified Euler's method find $y(0.2)$ correct to four decimal places solving the differential equation $\frac{dy}{dx} = x - y^2$, $y(0) = 1$. Taking $h = 0.1$.

   (c) Use Taylor's series method to find y at $x = 0.1,\ 0.2,\ 0.3$ (up to third derivatives) given $\frac{dy}{dx} = x^2 + y^2$ with $y(0) = 1$. Apply Milne's predictor-corrector formula to find $y(0.4)$. $(6 + 7 + 7 = 20)$

2. (a) Discuss the procedure for solving the second order differential equation $\frac{d^2y}{dx^2} = f\left(x, y, \frac{dy}{dx}\right)$ by Milne's method.

   (b) Use fourth order Runge-Kutta method to solve the system of equations $\frac{dy}{dx} = y + 2z$, $\frac{dz}{dx} = 3y + 2z$ given $y = 6$, $z = 4$ at $x = 0$ for y and z at $x = 0.02$

   (c) Use Picard's method to approximate y when $x = 0.1$ given that $\frac{d^2y}{dx^2} + 2x\frac{dy}{dx} + y = 0$ and $y = 0.5$, $\frac{dy}{dx} = 0.1$ when $x = 0$.

   $(4 + 8 + 8 = 20)$

3. (a) Define analytic function. Obtain Cauchy-Riemann equations in polar form as a set of necessary conditions for $f(z)$ to be analytic.

(b) Show that $u = e^{2x}(x \cos 2y - yhn\, 2y)$ is harmonic. Find the analytic function $f(z) = u + iv$.

(c) If $w = \phi + i\psi$ represents the complex potential for an electric field and $\psi = x^2 - y^2 + \dfrac{x}{x^2+y^2}$, determine the function ϕ.

$(7+7+6=20)$

4. (a) Find the bilinear transformation that maps the points $z = 1,\ i,\ -1$ into the points $w = 0,\ 1,\ \infty$

(b) Discuss the transformation $w = z + \dfrac{1}{z}$.

(c) Evaluate, using Cauchy's integral formula

$$\int_c \frac{e^z}{(z-1)(z+2)}\,dz, \text{ where } c \text{ is the circle } |z| = 3.$$

$(7+7+6=20)$

PART – B

5. (a) Obtain the solution of Laplace's equation $\dfrac{\partial^2 u}{\partial r^2} + \dfrac{1}{r}\dfrac{\partial u}{\partial r} + \dfrac{1}{r^2}\dfrac{\partial^2 u}{\partial \theta^2} + \dfrac{\partial^2 u}{\partial z^2} = 0$ in cylindrical coordinates leading to Legendre's equation.

(b) Obtain the series solution of Bessel's equation $x^2y'' + xy' + (x^2 - n^2)\,y = 0$ in the form $y = AJ_n(x) + BJ_{-n}(x)$.

(c) Prove the Rodrigue's formula $P_n(x) = \dfrac{1}{2^n n!}\dfrac{d^n}{dx^n}\left[(x^2-1)^n\right]$

$(7+7+6=20)$

6. (a) A can hit a target 3 times in 5 shots, B, 2 times in 5 shots and C 3 times in 4 shots. They fire a volley. What is the probability that: (i) two shots hit? (ii) At least two shots hit?

(b) The chance that a doctor A will diagnose a disease correctly is 60%. The chance that a patient of A will die after correct diagnosis is 40% and the chance of death by wrong diagnosis is 70%. If the patient of A dies what is the chance that his disease was correctly diagnosed?

(c) State and prove Baye's theorem.

$(7+7+6=20)$

7. (a) The probability distribution of a finite random variable X is given by the following table.

X_i:	-2	-1	0	1	2	3
$P(X_i)$:	0.1	K	0.2	$2K$	0.3	K

(i) Find the value of K (ii) Calculate mean and variance (iii) Evaluate $P(X \leq 1)$.

(b) Find the mean and variance of the binomial distribution.

(c) The length of a telephone conversation is an exponential variate with mean 3 minutes. Find the probability that a cell (i) ends in less than 3 minutes (ii) takes between 3 to 5 minutes.

$(7+7+6=20)$

8. (a) Explain the following terms:
(i) Null hypothesis (ii) Confidence limits (iii) Type I & II errors.

(b) A mechanist is making engine parts with axle diameter of 0.7 inches. A random sample of 10 parts showed a mean of 0.742 inches with standard deviation of 0.04 inches. On the basis of this sample, can it be concluded that the work is inferior at 5% level of significance.

(c) For the following data test the hypothesis that the accidents are uniformly distributed over all the days of the week for 99% confidence.

Day:	Sun	Mon	Tue	Wed	Thu	Fri	Sat	Total
No of accidents:	14	16	8	12	11	9	14	= 84

$(6+7+7=20)$

Table 1. Area under the normal curve from 0 to z

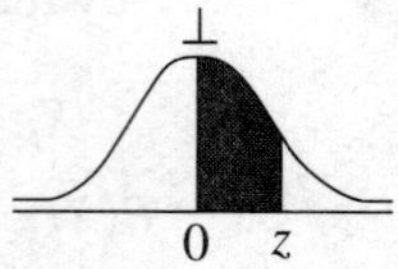

z	0	1	2	3	4	5	6	7	8	9
0.0	0.0000	0.0040	0.0080	0.0120	0.0160	0.0199	0.0239	0.0279	0.0319	0.0359
0.1	0.0398	0.0438	0.0478	0.0517	0.0557	0.0596	0.0636	0.0675	0.0714	0.0754
0.2	0.0793	0.0832	0.0871	0.0910	0.0948	0.0987	0.1026	0.1064	0.1103	0.1141
0.3	0.1179	0.1217	0.1255	0.1293	0.1331	0.1368	0.1406	0.1443	0.1480	0.1517
0.4	0.1554	0.1591	0.1628	0.1664	0.1700	0.1736	0.1772	0.1808	0.1844	0.1879
0.5	0.1915	0.1950	0.1985	0.2019	0.2054	0.2088	0.2123	0.2157	0.2190	0.2224
0.6	0.2258	0.2291	0.2324	0.2357	0.2389	0.2422	0.2454	0.2486	0.2518	0.2549
0.7	0.2580	0.2612	0.2642	0.2673	0.2704	0.2734	0.2764	0.2794	0.2823	0.2852
0.8	0.2881	0.2910	0.2939	0.2967	0.2996	0.3023	0.3051	0.3078	0.3106	0.3133
0.9	0.3159	0.3186	0.3212	0.3238	0.3264	0.3289	0.3315	0.3340	0.3365	0.3389
1.0	0.3413	0.3438	0.3461	0.3485	0.3508	0.3531	0.3554	0.3577	0.3599	0.3621
1.1	0.3643	0.3665	0.3686	0.3708	0.3729	0.3749	0.3770	0.3790	0.3810	0.3830
1.2	0.3849	0.3869	0.3888	0.3907	0.3925	0.3944	0.3962	0.3980	0.3997	0.4015
1.3	0.4032	0.4049	0.4066	0.4082	0.4099	0.4115	0.4131	0.4147	0.4162	0.4177
1.4	0.4192	0.4207	0.4222	0.4236	0.4251	0.4265	0.4279	0.4292	0.4306	0.4319
1.5	0.4332	0.4345	0.4357	0.4370	0.4382	0.4394	0.4406	0.4418	0.4429	0.4441
1.6	0.4452	0.4463	0.4474	0.4484	0.4495	0.4505	0.4515	0.4525	0.4535	0.4545
1.7	0.4554	0.4564	0.4573	0.4582	0.4591	0.4599	0.4608	0.4616	0.4625	0.4633
1.8	0.4641	0.4649	0.4656	0.4664	0.4671	0.4678	0.4686	0.4693	0.4699	0.4706
1.9	0.4713	0.4719	0.4726	0.4732	0.4738	0.4744	0.4750	0.4756	0.4761	0.4767
2.0	0.4772	0.4778	0.4783	0.4788	0.4793	0.4798	0.4803	0.4808	0.4812	0.4817
2.1	0.4821	0.4826	0.4830	0.4834	0.4838	0.4842	0.4846	0.4850	0.4854	0.4857
2.2	0.4861	0.4864	0.4868	0.4871	0.4875	0.4878	0.4881	0.4884	0.4887	0.4890
2.3	0.4893	0.4896	0.4898	0.4901	0.4904	0.4906	0.4909	0.4911	0.4913	0.4916
2.4	0.4918	0.4920	0.4922	0.4925	0.4927	0.4929	0.4931	0.4932	0.4934	0.4936
2.5	0.4938	0.4940	0.4941	0.4943	0.4945	0.4946	0.4948	0.4949	0.4951	0.4952
2.6	0.4953	0.4955	0.4956	0.4957	0.4959	0.4960	0.4961	0.4962	0.4963	0.4964
2.7	0.4965	0.4966	0.4967	0.4968	0.4969	0.4970	0.4971	0.4972	0.4973	0.4974
2.8	0.4974	0.4975	0.4976	0.4977	0.4977	0.4978	0.4979	0.4979	0.4980	0.4981
2.9	0.4981	0.4982	0.4982	0.4983	0.4984	0.4984	0.4985	0.4985	0.4986	0.4986
3.0	0.4987	0.4987	0.4987	0.4988	0.4988	0.4989	0.4989	0.4989	0.4990	0.4990
3.1	0.4990	0.4991	0.4991	0.4991	0.4992	0.4992	0.4992	0.4992	0.4993	0.4993

Table 2. Values of [*t*] with probability *P* and degrees of freedom *v*

v \ *P*	*0.50*	*0.10*	*0.05*	*0.02*	*0.01*
1	1.000	6.34	12.71	31.82	63.66
2	0.816	2.92	4.30	6.96	9.92
3	0.765	2.35	3.18	4.54	5.84
4	0.741	2.13	2.78	3.75	4.60
5	0.727	2.02	2.57	3.36	4.03
6	0.718	1.94	2.45	3.14	3.71
7	0.711	1.90	2.36	3.00	3.50
8	0.706	1.86	2.31	2.90	3.36
9	0.703	1.83	2.26	2.82	3.25
10	0.700	1.81	2.23	2.76	3.17
11	0.697	1.80	2.20	2.72	3.11
12	0.695	1.78	2.18	2.68	3.06
13	0.694	1.77	2.16	2.65	3.01
14	0.692	1.76	2.14	2.62	2.98
15	0.691	1.75	2.13	2.60	2.95
16	0.690	1.75	2.12	2.58	2.92
17	0.689	1.74	2.11	2.57	2.90
18	0.688	1.73	2.10	2.55	2.88
19	0.688	1.73	2.09	2.54	2.86
20	0.687	1.72	2.09	2.53	2.84
21	0.686	1.72	2.08	2.52	2.83
22	0.686	1.72	2.07	2.51	2.82
23	0.685	1.71	2.07	2.50	2.81
24	0.685	1.71	2.06	2.49	2.80
25	0.684	1.71	2.06	2.48	2.79
26	0.684	1.71	2.06	2.48	2.78
27	0.684	1.70	2.05	2.47	2.77
28	0.683	1.70	2.05	2.47	2.76
29	0.683	1.70	2.04	2.46	2.76
30	0.683	1.70	2.04	2.46	2.75

Table 3. Values of χ^2 with probability P and df v

v \ P	0.99	0.95	0.50	0.30	0.20	0.10	0.05	0.01
1	0.0002	0.004	0.46	1.07	1.64	2.71	3.84	6.64
2	0.020	0.103	1.39	2.41	3.22	4.60	5.99	9.21
3	0.115	0.35	2.37	3.66	4.64	6.25	7.82	11.34
4	0.30	0.71	3.36	4.88	5.99	7.78	9.49	13.28
5	0.55	1.14	4.35	6.06	7.29	9.24	11.07	15.09
6	0.87	1.64	5.35	7.23	8.56	10.64	12.59	16.81
7	1.24	2.17	6.35	8.38	9.80	12.02	14.07	18.48
8	1.65	2.73	7.34	9.52	11.03	13.36	15.51	20.09
9	2.09	3.32	8.34	10.66	12.24	14.68	16.92	21.67
10	2.56	3.94	9.34	11.78	13.44	15.99	18.31	23.21
11	3.05	4.58	10.34	12.90	14.63	17.28	19.68	24.72
12	3.57	5.23	11.34	14.01	15.81	18.55	21.03	26.22
13	4.11	5.89	12.34	15.12	16.98	19.81	22.36	27.69
14	4.66	6.57	13.34	16.22	18.15	21.06	23.68	29.14
15	5.23	7.26	14.34	17.32	19.31	22.31	25.00	30.58
16	5.81	7.96	15.34	18.42	20.46	23.54	26.30	32.00
17	6.41	8.67	16.34	19.51	21.62	24.77	27.59	33.41
18	7.02	9.39	17.34	20.60	22.76	25.99	28.87	34.80
19	7.63	10.12	18.34	21.69	23.90	27.20	30.14	36.19
20	8.26	10.85	19.34	22.78	25.04	28.41	31.41	37.57
21	8.90	11.59	20.34	23.86	26.17	29.62	32.67	38.93
22	9.54	12.34	21.34	24.94	27.30	30.81	33.92	40.29
23	10.20	13.09	22.34	26.02	28.43	32.01	35.17	41.64
24	10.86	13.85	23.34	27.10	29.55	33.20	36.42	42.98
25	11.52	14.61	24.34	28.17	30.68	34.68	37.65	44.31
26	12.20	15.38	25.34	29.25	31.80	35.56	38.88	45.64
27	12.88	16.15	26.34	30.32	32.91	36.74	40.11	46.96
28	13.56	16.93	27.34	31.39	34.03	37.92	41.34	48.28
29	14.26	17.71	28.34	32.46	35.14	39.09	42.56	49.59
30	14.95	18.49	29.34	33.53	36.25	40.26	43.77	50.89

Table 4. 5% and 1% points of F

ν_1 / ν_2	1	2	3	4	5	6	8	12	24	∞
2	18.51	19.00	19.16	19.25	19.30	19.32	19.37	19.41	19.45	19.50
	98.49	99.00	99.17	99.25	99.30	99.33	99.36	99.42	99.46	99.50
3	10.13	9.55	9.28	9.12	9.01	8.94	8.84	8.74	8.64	8.53
	34.12	30.82	29.46	28.71	28.24	27.91	27.49	27.05	26.60	26.12
4	7.71	6.94	6.59	6.39	6.26	6.16	6.04	5.91	5.77	5.63
	21.20	18.00	16.69	15.98	15.52	15.21	14.80	14.37	13.93	13.46
5	6.61	5.79	5.41	5.19	5.05	4.95	4.82	4.68	4.53	4.36
	16.26	13.27	12.06	11.39	10.97	10.67	10.27	9.89	9.47	9.02
6	5.99	5.14	4.76	4.53	4.39	4.28	4.15	4.00	3.84	3.67
	13.74	10.92	9.78	9.15	8.75	8.47	8.10	7.72	7.31	6.88
7	5.59	4.74	4.35	4.12	3.97	3.87	3.73	3.57	3.41	3.23
	12.25	9.55	8.45	7.85	7.46	7.19	6.84	6.47	6.07	5.65
8	5.32	4.46	4.07	3.84	3.69	3.58	3.44	3.28	3.12	2.93
	11.26	8.65	7.59	7.01	6.63	6.37	6.03	5.67	5.28	4.86
9	5.12	4.26	3.86	3.63	3.48	3.37	3.23	3.07	2.90	2.71
	10.56	8.02	6.99	6.42	6.06	5.80	5.47	5.11	4.73	4.31
10	4.96	4.10	3.71	3.48	3.33	3.22	3.07	2.91	2.74	2.54
	10.04	7.56	6.55	5.99	5.64	5.39	5.06	4.71	4.33	3.91
12	4.75	3.88	3.49	3.26	3.11	3.00	2.85	2.69	2.50	2.30
	9.33	6.93	5.95	5.41	5.06	4.82	4.50	4.16	3.78	3.36
14	4.60	3.74	3.34	3.11	2.96	2.85	2.70	2.53	2.35	2.13
	8.86	6.51	5.56	5.03	4.69	4.46	4.14	3.80	3.43	3.00
16	4.49	3.63	3.24	3.01	2.85	2.74	2.59	2.42	2.24	2.01
	8.53	6.23	5.29	4.77	4.44	4.20	3.89	3.55	3.18	2.75
18	4.41	3.55	3.16	2.93	2.77	2.66	2.51	2.34	2.15	1.92
	8.28	6.01	5.09	4.58	4.25	4.01	3.71	3.37	3.01	2.57
20	4.35	3.49	3.10	2.87	2.71	2.60	2.45	2.28	2.08	1.84
	8.10	5.85	4.94	4.43	4.10	3.87	3.56	3.23	2.86	2.42
25	4.24	3.38	2.99	2.76	2.60	2.49	2.34	2.16	1.96	1.71
	7.77	5.57	4.68	4.18	3.86	3.63	3.32	2.99	2.62	2.17
30	4.17	3.32	2.92	2.69	2.53	2.42	2.27	2.09	1.89	1.62
	7.56	5.39	4.51	4.02	3.70	3.47	3.17	2.84	2.47	2.01
40	4.08	3.23	2.84	2.61	2.45	2.34	2.18	2.00	1.79	1.51
	7.31	5.18	4.31	3.83	3.51	3.29	2.99	2.66	2.29	1.81
60	4.00	3.15	2.76	2.52	2.37	2.25	2.10	1.92	1.70	1.39
	7.08	4.98	4.13	3.65	3.34	3.12	2.82	2.50	2.12	1.60

Index